T0269081

LONDON MATHEMATICAL SOCIETY LECTURE NOTE SERIES

Managing Editor: Professor J.W.S. Cassels, Department of Pure Mathematics and Mathematical Statistics, University of Cambridge, 16 Mill Lane, Cambridge CB2 1SB, England

The titles below are available from booksellers, or, in case of difficulty, from Cambridge University Press.

London Mathematical Society Lecture Note Series. 217

Quadratic forms with applications to algebraic geometry and topology

Albrecht Pfister
Johannes Gutenberg-Universität, Mainz

CAMBRIDGE
UNIVERSITY PRESS

Published by the Press Syndicate of the University of Cambridge
The Pitt Building, Trumpington Street, Cambridge CB2 1RP
40 West 20th Street, New York, NY 10011-4211, USA
10 Stamford Road, Oakleigh, Melbourne 3166, Australia

First published 1995

Library of Congress cataloging in publication data

Pfister, Albrecht
Quadratic forms with applications to algebraic geometry and topology / Albrecht Pfister
   p. cm. -- (London Mathematical Society lecture note series; 217)
Includes bibliographical references (p. -) and index.
ISBN 0-521-46755-1 (paperback)
1. Forms, Quadratic. 2. Geometry, Algebraic. 3. Topology.
I. Title. II Series.
QA243.P45 1996
512'.74--dc20   95-13802   CIP

British Library cataloguing in publication data available

ISBN 0 521 46755 1 paperback

Transferred to digital printing 2003

Contents

Preface

This book grew out of a graduate course that I gave at the University of Cambridge in the Easter Term of 1993. The idea of publishing a somewhat enlarged and polished version of my lectures came from Professor J.W.S. Cassels, who, in addition, made it possible for me to spend my sabbatical at Cambridge supported by a Research Grant of the SERC and an appointment as a "Visiting Fellow Commoner" of Trinity College. I thank these institutions for their help in making my stay in Cambridge a very pleasant one.

I should point out in this connection that a great deal of my research on quadratic forms began in the year 1963 when I attented a colloquium talk given by Cassels on "Sums of Squares of Rational Functions" at the University of Göttingen. Later, our connections intensified during the Academic Year 1966/67 when I studied and lectured in Cambridge. My early Lecture Notes [Pfister 1967$_1$] give an idea of the status of the algebraic theory of quadratic forms in those days. Thus, much of my previous work as well as the present book owe their existence to the constant encouragement and interest of Cassels over many years. For this reason, I wish to express my deep gratitude to him.

This book is not a systematic treatise on quadratic forms. Excellent books of this kind are already available, in particular the books of O'Meara [O'M] on the arithmetic theory over number fields and their integer domains and the books of Lam [L] and Scharlau [S] on the algebraic theory over general fields.

The choice of material considered herein reflects my own interests and incorporates a considerable amount of my scientific work over the past 30 years. It starts with some "highlights" about quadratic forms in Chapters 1 and 2. A main theme of the text concerns the field invariants: "level" (Chapter 3), "Pythagoras number" (Chapter 7), and "u-invariant" (Chapter 8). Many people contributed to the results presented here. Furthermore, I have emphasized the way in which quadratic forms lead to rich interconnections linking algebra, number theory, algebraic geometry, and algebraic topology. Such topics are covered in Chapters 3, 4, 5, 6 and 10. Finally, systems of quadratic forms (Chapter 9) serve as a kind of clue for relating algebraic geometry and topology to quadratic forms. The specific topics of the various sections can best be seen from the table of contents, and so there is no point in repeating them here.

The prerequisites on the part of the reader are fairly modest. Standard knowledge from introductory courses suffices for most parts of the text. In several places where I need more advanced results a precise reference is given. I have tried to make the main body of the book self-contained with full proofs. Side results or more difficult theorems which go far beyond the methods used here are given without proofs. Examples, notes and open questions have been added whenever possible. They can be used by the reader both to clarify his understanding and to extend his knowledge of the concepts.

I hope that the book will prove equally well suited for graduate students, teachers, researchers on quadratic forms, and mathematicians working in other disciplines with an interest in the topics treated here. My special thanks go to Michael Meurer for proof-reading the manuscript and to Mrs Jutta Gonska for preparing an excellent typescript.

Mainz, December 1994   Albrecht Pfister

Chapter 1

The Representation Theorems of Cassels

§1. Preliminaries on Quadratic Forms

1.1 Definition. Let K be a (commutative) field, let n be a natural number. An n-ary *quadratic form* over K is a homogeneous polynomial of degree 2 in n variables with coefficients from K. It has the form

$$\varphi(x_1,\ldots,x_n) = \sum_{i,j=1}^{n} a_{ij}x_i x_j \in K[x_1,\ldots,x_n].$$

In matrix notation this can be written as follows:
Let x be the column vector with components $x_1,\ldots,x_n$, let x' be its transpose which is a row vector and let $A = (a_{ij})$ be the (n by n)-matrix in $M_{n,n}(K)$ which is determined by the coefficients a_{ij} of φ. Then

$$\varphi(x) = x'Ax.$$

1.2 Definition. Two n-ary quadratic forms φ and ψ over K are called *equivalent* if there is a nonsingular linear transformation $T \in GL_n(K)$ such that

$$\psi(x) = \varphi(Tx).$$

Clearly this is an equivalence relation. We write

$$\psi \cong \varphi \quad (\text{over } K).$$

From now on we shall suppose that the characteristic of K is different from 2. The case char $K = 2$ is postponed to section 4.

For char $K \neq 2$ we can replace the coefficients a_{ij} by $\frac{a_{ij}+a_{ji}}{2}$ without changing the quadratic form φ. Then $a_{ij} = a_{ji}$, i.e. A is *symmetric*. Under an equivalence T the symmetric matrix A is replaced by the congruent matrix

$$B = T'AT$$

which is again symmetric. Furthermore, we see that the polynomial

$$\varphi(x) = \sum_i a_{ii}x_i^2 + \sum_{i<j}(a_{ij} + a_{ji})x_i x_j$$

uniquely determines the matrix A if A is symmetric since the a_{ii} and $2a_{ij}$ (for $i < j$) are exactly the coefficients of φ.

1.3 Definition. Let char $K \neq 2$, let $\varphi(x) = x'Ax$ be a quadratic form over K with $A = A'$ and let x, y be independent indeterminate vectors. Put

$$b_\varphi(x, y) = \frac{1}{2}(\varphi(x + y) - \varphi(x) - \varphi(y)) = x'Ay = y'Ax.$$

b_φ is called "the associated symmetric bilinear form" of φ.

Conversely, any symmetric bilinear form

$$b(x, y) = x'Ay \qquad \text{with} \quad A = A'$$

determines a quadratic form $\varphi(x) := b(x, x)$, and these two processes are inverse to one another. Therefore the theories of quadratic forms over K and of symmetric bilinear forms over K (in finitely many variables) essentially coincide if char $K \neq 2$.

Every n-ary quadratic form φ over K induces a map Q_φ from the vector-space $V = K^n$ of n-fold column vectors over K to the field K, namely

$$Q_\varphi : V \to K, \quad Q_\varphi(v) := \varphi(v) \quad \text{for } v \in V.$$

Q_φ is a *quadratic map*, i.e. it has the following properties:

(1) $Q_\varphi(av) = a^2 Q_\varphi(v)$ for $a \in K$, $v \in V$.

(2) The map $B_\varphi : V \times V \to K$ given by

$$B_\varphi(v, w) = \frac{1}{2}(Q_\varphi(v + w) - Q_\varphi(v) - Q_\varphi(w))$$

is K-bilinear (and symmetric).

If $\varphi(x) = x'Ax$ is given by the symmetric matrix $A = (a_{ij})$ and if $e_1, \ldots, e_n$ is the standard basis of V then

$$Q_\varphi(e_i) = a_{ii} \qquad \text{and} \qquad B_\varphi(e_i, e_j) = \frac{1}{2}(a_{ij} + a_{ji}) = a_{ij}.$$

This means that A and φ can be reconstructed from the pair (Q_φ, B_φ).

This observation leads to the following definitions and proposition.

1.4 Definition. Let V be an n-dimensional K-vector-space. A map $Q : V \to K$ is called a *quadratic map* and the pair (V, Q) is then called a *quadratic space* over K if Q satisfies the conditions:

(1) $Q(av) = a^2 Q(v)$ for $a \in K, v \in V$.

(2) The map $B : V \times V \to K$ given by

$$B(v, w) := \frac{1}{2}(Q(v + w) - Q(v) - Q(w))$$

is K-bilinear.

1.5 Definition. Two n-dimensional quadratic spaces (V, Q) and (V', Q') over K are called *isometric* if there exists a K-linear isomorphism $T : V \rightarrow V'$ such that

$$Q(v) = Q'(Tv) \quad \text{for all } v \in V.$$

We write: $(V, Q) \cong (V', Q')$.

1.6 Proposition. There is a 1-1-correspondence between equivalence classes of n-ary quadratic forms over K and isometry classes of n-dimensional quadratic spaces over K.

PROOF. The correspondence $\varphi \rightsquigarrow Q_\varphi$ constructed above for $V = K^n$ has the desired properties since φ can be regained from Q_φ and since every n-dimensional K-vector-space V is isomorphic to K^n.

This enables us to switch from the more algebraic language of quadratic forms to the more geometric language of quadratic spaces and vice versa. The latter point of view was introduced in the fundamental paper [Witt 1937] of Witt and has been proved very useful. If there is no danger of confusion we will no longer distinguish between the form φ and the map Q_φ, i.e. we write φ instead of Q_φ and b_φ instead of B_φ.

1.7 Orthogonal Sums. Two quadratic spaces (V_1, φ_1) and (V_2, φ_2) over K of dimensions n_1 and n_2 respectively, give rise to a quadratic space (V, φ) of dimension $n = n_1 + n_2$, namely

$$\begin{aligned} V &= V_1 \oplus V_2, \\ \varphi(v) &= \varphi(v_1) + \varphi(v_2), \end{aligned}$$

for $v_1 \in V_1$, $v_2 \in V_2$ and $v = v_1 + v_2 \in V$. This space (V, φ) is called the orthogonal sum of (V_1, φ_1) and (V_2, φ_2). We also write $\varphi = \varphi_1 \oplus \varphi_2$. If φ_i is given by the symmetric matrix A_i $(i = 1, 2)$ then φ has matrix

$$A = \begin{pmatrix} A_1 & 0 \\ 0 & A_2 \end{pmatrix}.$$

Similarly, the orthogonal sum of r quadratic spaces can be defined for any $r \in \mathbb{N}$. Up to equivalence it depends only on (the equivalence classes of) the summands but not on their order.

Conversely, let (V, φ) be a quadratic space and let V_i $(i = 1, \ldots, r)$ be subspaces of V such that $V = V_1 \oplus \ldots \oplus V_r$ and $b_\varphi(v_i, v_j) = 0$ for $v_i \in V_i$, $v_j \in V_j$, $i \neq j$. Then $\varphi = \varphi_1 \oplus \ldots \oplus \varphi_r$ with $\varphi_i = \varphi|_{V_i}$, i.e. φ is the orthogonal sum of the forms φ_i.

We can now prove

1.8 Theorem. Let char $K \neq 2$. Then every quadratic space (V, φ) over K is isometric to an orthogonal sum of 1-dimensional spaces. In other words: Every n-ary quadratic form φ over K is equivalent to a *diagonal form* ψ with $\psi(x) = \sum_1^n a_i x_i^2$, $a_i \in K$.

PROOF. We use induction on $\dim V = n$. If $\varphi(v) = 0$ for all $v \in V$ then 1.3 shows $b_\varphi(v_1, v_2) = 0$ for any pair $v_1, v_2 \in V$. In this case any basis $\{v_1, \ldots, v_n\}$ of V is an orthogonal basis.

If $\varphi(v_1) = a_1 \neq 0$ for some $v_1 \in V$ we consider the subspace

$$U = (Kv_1)^\perp = \{u \in V : b_\varphi(u, v_1) = 0\}$$

of all vectors u which are orthogonal to v_1 with respect to b_φ. The condition $b_\varphi(u, v_1) = 0$ amounts to one linear equation for u. Since $\varphi(v_1) = b_\varphi(v_1, v_1) \neq 0$ we have $v_1 \notin U$ and $\dim U = n - 1$. This shows $V = Kv_1 \oplus U$ and $\varphi = \varphi_1 \oplus \varphi_2$ with $\varphi_1 = \varphi|_{Kv_1}$, $\varphi_2 = \varphi|_U$. The induction hypothesis for U finishes the proof.

Note. In the case $\varphi \neq 0$ the element $a_1 \in K^\bullet = K\backslash\{0\}$ is any element which has the form $\varphi(v_1)$, $v_1 \in V$.

Notation. The diagonal form $\psi(x) = \sum_1^n a_i x_i^2$ is abbreviated by

$$\psi = \langle a_1, \ldots, a_n \rangle = \langle a_1 \rangle \oplus \ldots \oplus \langle a_n \rangle.$$

1.9 Definition. Let $A = A'$ be a symmetric matrix. Let (V, φ) with $\varphi(x) = x'Ax$ be the corresponding quadratic space.

(1) The subspace $\mathrm{rad}\, V = V^\perp = \{u \in V : b_\varphi(u, v) = 0 \text{ for all } v \in V\}$ is called the *radical of* (V, φ), and is written as $\mathrm{rad}\, V$.

(2) (V, φ) is called *regular* if $\mathrm{rad}\, V = 0$.

The following observations are immediate:

- $\mathrm{rad}\, V = \{u \in V : u'Av = 0 \text{ for all } v \in V\} = \{u \in V : u'A = 0\}$.

- $\mathrm{rad}\, V = 0 \iff \det A \neq 0$.

- The terms *radical* and *regular* are invariant under isometry.

- If φ is not regular then $\varphi \cong \langle a_1, \ldots a_n \rangle$ and, say, $a_n = 0$.

This means that φ can be transformed into a quadratic form which actually depends on at most $n - 1$ variables. Since n can be any natural number in our treatment of quadratic forms we can and will henceforth assume that all forms are regular.

Note. Let φ be a quadratic form over K and let $L \supset K$ be any extension field of K. Then φ may also be considered as a quadratic form over L. This "extended" form is usually denoted by φ_L or $\varphi \otimes L$. We have

$$\varphi = \varphi_K \text{ regular} \iff \varphi_L \text{ regular}.$$

1.10 Definition. For an n-ary quadratic form φ over K we introduce the following notions:

(1) For $a \in K$ we say that "φ *represents* a over K", if there is a *nonzero* vector $0 \neq v \in K^n$ such that

$$\varphi(v) = a.$$

(2) $D_K(\varphi) = \{\varphi(v) : 0 \neq v \in K^n\}$ is the set of all those elements of K which are represented by φ over K.

(3) $D_K^\bullet(\varphi) = D_K(\varphi)\backslash\{0\} \subseteq K^\bullet$.

(4) φ is called *universal* (over K) if $D_K(\varphi) = K^\bullet$.

(5) φ is called *isotropic* (over K) if $0 \in D_K(\varphi)$, otherwise φ is called *anisotropic* (over K).

Example. Consider the form $\varphi = \langle 1, 1 \rangle$, i.e. $\varphi(x) = x_1^2 + x_2^2$ over the fields **R** and **C**:
Over **R** φ does not represent the elements -1 and 0 since $r_1^2 + r_2^2 > 0$ for any pair $(r_1, r_2) \neq (0, 0)$ of real numbers.
Over **C** φ does represent -1 and 0 since $-1 = i^2$, $0 = 1^2 + i^2$. Furthermore, φ is universal over **C**.
This shows that the notions of Definition 1.10 depend very much on the field K, not only on φ.
Clearly a 1-dimensional regular space $\varphi = \langle a \rangle, a \in K^\bullet$, can never be isotropic. Let us study the 2-dimensional regular isotropic spaces over K.

1.11 Proposition. Up to equivalence there is just one regular isotropic quadratic form φ of dimension 2, namely $\varphi(x) = 2x_1 x_2$. We have

$$\varphi \cong \langle a, -a \rangle$$

for an arbitrary $a \in K^\bullet$. In particular φ is universal.

PROOF. Let $0 \neq v_1 \in V = K^2$ be an isotropic vector. Since φ is regular there exists $u \in V$ such that $b_\varphi(v_1, u) \neq 0$ and by multiplying u by a suitable element of $K^\bullet$ we can arrange $b_\varphi(v_1, u) = 1$. Clearly u is K-linearly independent from v_1 since $b_\varphi(v_1, v_1) = \varphi(v_1) = 0$. For any $\lambda \in K$ the vectors v_1 and $v_2 = u + \lambda v_1$ form a basis of V for which $\varphi(v_1) = 0$ and $b_\varphi(v_1, v_2) = 1$. Finally, $\varphi(v_2) = \varphi(u) + 2\lambda b_\varphi(u, v_1) + \lambda^2 \varphi(v_1) = \varphi(u) + 2\lambda$. Choosing $\lambda = -\frac{\varphi(u)}{2}$ we get $\varphi(v_2) = 0$.

For an indeterminate vector $x = x_1 v_1 + x_2 v_2$ this gives

$$\varphi(x) = x_1^2 \varphi(v_1) + 2x_1 x_2 b_\varphi(v_1, v_2) + x_2^2 \varphi(v_2) = 2x_1 x_2.$$

For any $a \in K^\bullet$ φ represents a: Take e.g. $x_1 = \frac{1}{2}$, $x_2 = a$. By Theorem 1.8 we get $\varphi \cong \langle a, a_2 \rangle$ for some $a_2 \in K^\bullet$. But φ is isotropic, hence $ac_1^2 + a_2 c_2^2 = 0$ for some pair $(c_1, c_2) \neq (0, 0)$ in K^2. Then $c_1 c_2 \neq 0$ and $a_2 = -a(\frac{c_1}{c_2})^2$. Therefore $\varphi \cong \langle a, -a \rangle$ because the coefficients in a diagonal matrix for φ can be multiplied by arbitrary nonzero squares from K without changing the equivalence class of φ.

Notation. The (equivalence class of a) regular isotropic quadratic form of dimension 2 over K is denoted by H. In other words: $H \cong \langle 1, -1 \rangle$. H is called the *hyperbolic plane*.

Proposition 1.11 can be generalized as follows:

1.12 Proposition. Let (V, φ) be a regular isotropic quadratic space over K with $\dim V = n \geq 2$. Then $V = U \oplus W$ with $U \cong H$, $\dim W = n - 2$; $\varphi \cong \langle 1, -1 \rangle \oplus \psi$ with $\psi = \varphi|_W$.

PROOF. As in 1.11 we find vectors $v_1, v_2 \in V$ such that the 2-dimensional subspace $U = Kv_1 + Kv_2$ of V together with the quadratic form $\varphi|_U$ is (isometric to) the hyperbolic plane H. Put $W = U^\perp = \{w \in V : b_\varphi(U, w) = 0\}$. Clearly $\dim W \geq n - 2$. On the other hand $U \cap U^\perp = \operatorname{rad} U = 0$ since $(U, \varphi|_U) \cong H$ is regular. Therefore $\dim W = n - 2$ and $V = U \oplus W$ (orthogonal sum). For the form φ this means $\varphi \cong \langle 1, -1 \rangle \oplus \psi$.

§2. The Main Theorem

We start with a simple observation. Let $\varphi(x) = \sum_{i,j=1}^n a_{ij} x_i x_j \in K[x_1, \ldots x_n]$ be a quadratic form over a field K. Let $L = K(t)$ be the *rational function field over K* in one variable t. Then we have

2.1 Lemma. φ anisotropic over K $\Rightarrow$ φ_L anisotropic over L.

PROOF. Assume $\varphi(f) = 0$ with $0 \neq f = (f_1, \ldots, f_n)$, $f_i \in L$. Choose a common denominator g_0 of the rational functions f_i. Then $f_i = \frac{g_i}{g_0}$ with $g_0, g_1, \ldots, g_n \in K[t]$ and $\varphi(g) = g_0^2 \varphi(f) = 0$ for $0 \neq g = (g_1, \ldots, g_n)$. Let now

$0 \neq d \in K[t]$ be the greatest common divisor of the polynomials $g_1, \ldots, g_n$. Then $g_i = dh_i$ with $h_i \in K[t]$, and $h_1, \ldots, h_n$ are relatively prime. Put $h = (h_1, \ldots, h_n)$. Then $\varphi(g) = d^2\varphi(h) = 0$ is an identity in t. Since $K[t]$ is an integral domain and $d = d(t) \neq 0$ we get $\varphi(h) = 0$. Put $c_i = h_i(0) \in K$, $c = (c_1, \ldots, c_n)$. The elements are not all zero since otherwise the $h_i(t)$ would all be divisible by t. Hence $0 \neq c \in K^n$ and $\varphi(c) = 0$ by substituting $t \to 0$ in the identity $\varphi(h) = 0$ in $K[t]$. This contradicts the anisotropy of φ.

2.2 Theorem. Let $\varphi(x) = \varphi(x_1, \ldots, x_n) = \sum_{i,j=1}^n a_{ij} x_i x_j$ be an n-ary quadratic form over the field K, char $K \neq 2$. Let $0 \neq p(t) \in K[t]$ be a polynomial in one variable. Suppose that φ represents $p = p(t)$ over the field $L = K(t)$. Then φ represents p over the ring $K[t]$, i.e. there are polynomials $f_i = f_i(t) \in K[t]$ such that $\varphi(f_1, \ldots, f_n) = p$.

PROOF.
1) If φ is not regular we may replace φ by a quadratic form in less than n variables and argue by induction on n. For $n = 1$, $\varphi(x) = a_{11} x_1^2$, $a_{11} f_1^2 = p$ with $f_1 \in K(t)$, the theorem is true since $f_1 \in K[t]$ follows automatically. (Use that $K[t]$ is a unique factorization domain.)
2) Suppose now that φ is regular but isotropic. Then $\varphi \cong H \oplus \psi$ over K by Proposition 1.12, i.e. without loss of generality

$$\varphi(x) = 2x_1 x_2 + \psi(x_3, \ldots, x_n).$$

Put $x_1 = p(t)$, $x_2 = \frac{1}{2}$, $x_3 = \ldots = x_n = 0$. This shows that φ represents p over $K[t]$.
3) From now on φ is (regular and) anisotropic. By assumption we have a representation

$$(1) \qquad \varphi\left(\frac{f_1}{f_0}, \ldots, \frac{f_n}{f_0}\right) = p$$

with polynomials $f_0, \ldots, f_n \in K[t]$. Without loss of generality the greatest common divisor of $f_0, \ldots, f_n$ is 1.

Furthermore we may suppose that under all representations of shape (1) the given one has *minimal* degree $d = \deg f_0 \geq 0$ of the denominator f_0. If $d = 0$ then f_0 is a nonzero constant and we are finished.

Hypothesis: $d > 0$.
Then we have to derive a contradiction. We introduce the $(n+1)$-dimensional quadratic form

$$(2) \qquad \psi = \langle -p(t) \rangle \oplus \varphi_L \qquad \text{over} \quad L = K(t).$$

Explicitly: $\psi(x_0, \ldots, x_n) = -p(t)x_0^2 + \varphi(x_1, \ldots, x_n)$.
(1) implies $\psi(f_0, \ldots, f_n) = 0$.

Apply the euclidean algorithm (division by f_0) to the polynomials f_i ($i = 0, \ldots, n$). This gives

(3) $f_i = f_0 g_i + r_i$ ($i = 0, \ldots, n$) with $g_i, r_i \in K[t]$, $\deg r_i < d$.

In particular, $g_0 = 1$, $r_0 = 0$, $\deg r_0 = -\infty$. Put $f = (f_0, \ldots, f_n)$, $g = (g_0, \ldots, g_n)$. Then $\psi(f) = 0$ and $\psi(g) \neq 0$ by the minimality condition on f_0 since $0 = \deg g_0 < \deg f_0 = d$. In particular, the nonzero vectors f and g are linearly independent over L.

(4) Define $h = \lambda f - \mu g \in L^{n+1}$ with $\lambda = \psi(g)$, $\mu = 2b_\psi(f, g)$.

We have $h = (h_0, \ldots, h_n)$, $h_i \in K[t]$. $\lambda \neq 0$ implies $h \neq 0$. On the other hand we get

(5) $\psi(h) = \lambda^2 \psi(f) - 2\lambda\mu b_\psi(f, g) + \mu^2 \psi(g) = \lambda^2 \cdot 0 - \lambda\mu^2 + \mu^2\lambda = 0.$

Actually we must have $h_0 \neq 0$. Otherwise $h = (0, h_1, \ldots, h_n) \neq 0$ would give a nontrivial solution of the equation

$$\psi(h) = \varphi(h_1, \ldots, h_n) = 0 \quad \text{over the field } L = K(t)$$

whereas φ is anisotropic over L by Lemma 2.1. It remains to estimate $\deg h_0$. We have

(6) $h_0 = \lambda f_0 - \mu = \psi(g)f_0 - 2b_\psi(f, g) = \dfrac{1}{f_0}\psi(f_0 g - f)$

$$= \frac{1}{f_0} \sum_{i,j=1}^{n} a_{ij}(f_0 g_i - f_i)(f_0 g_j - f_j).$$

This implies

$$\deg \psi(f_0 g - f) \leq 2 \max_{i=1,\ldots,n} \deg(f_0 g_i - f_i) = 2 \max_{i=1,\ldots,n} \deg r_i \leq 2(d-1),$$

hence

(7) $\deg h_0 = -d + \deg \psi(f_0 g - f) \leq d - 2.$

Thus h would give a solution of (1) which is "smaller" than f: Contradiction. The proof of 2.2 is finished.

Note. The geometric idea behind the proof of 2.2 is as follows: The equation $\psi = 0$ defines a quadric (hypersurface of degree 2) Q in the projective n-space over L. The "points" f, g are different with $f \in Q$, $g \notin Q$. The "line" joining f and g intersects Q in a second point $h \neq f$. It turns out that the choice (3) for g leads to $\deg h_0 < \deg f_0$.

Theorem 2.2 has the following partial generalization.

2.3 Generalization. Let $\varphi(x) = \sum_{i,j=1}^{n} a_{ij}x_i x_j$ be a quadratic form over $L = K(t)$ such that $a_{ij} \in K[t]$ and $\deg a_{ij} \leq 1$ for all (i, j). Suppose φ is *anisotropic* over L. Let $0 \neq p(t) \in K[t]$ be a polynomial which is represented by φ over L. Then p is already represented over $K[t]$.

PROOF. Part 3) of the above proof carries over verbatim to this slightly more general case. The only change is

$$\deg \psi(f_0 g - f) \leq 1 + 2 \max \deg r_i \leq 2d - 1,$$

hence

(7')
$$\deg h_0 \leq d - 1 < d.$$

This is still enough to derive the contradiction.

Note. The generalization 2.3 is no longer valid if φ is isotropic. Let $\varphi = \langle t, -t \rangle$, $p(t) = 1$. φ is clearly isotropic, hence universal over $L = K(t)$. Thus φ represents $p = 1$ over L. (Derive such a representation explicitly!) But there is clearly no solution of $t f_1^2 - t f_2^2 = 1$ with polynomials $f_1, f_2 \in K[t]$.

Note. At first sight it seems that repeated application of Theorem 2.2 would give the corresponding result for a polynomial $p = p(t_1, \ldots, t_r)$ in several variables. But a closer look reveals that starting from a representation of $p(t_1, t_2)$ over the ring $K(t_2)[t_1]$ the procedure of the above proof with respect to the variable t_2 leads to a representation over $K(t_1)[t_2]$ and not over $K[t_1][t_2]$ since $K[t_1, t_2]$ is no longer a euclidean domain. Actually the *existence* of counter-examples over $\mathbf{R}(t_1, t_2)$ for $\varphi = \underbrace{\langle 1, \ldots, 1 \rangle}_{n}$ with suitable n goes far back to Hilbert[1888]. Nevertheless the first explicit counter-example (for $n = 4$, $r = 2$) was only found in the year 1967 by Motzkin[1967]. It reads as follows:

2.4 Example. Let $p(x, y) = 1 - 3x^2 y^2 + x^4 y^2 + x^2 y^4 \in \mathbf{R}[x, y]$. Then

(1) p is a sum of four squares in the ring $\mathbf{R}(x)[y]$, hence also in the field $\mathbf{R}(x, y)$.

(2) p is not a sum of (any finite number of) squares in the polynomial ring $\mathbf{R}[x, y]$.

PROOF. 1) Check the following identities:

$$p(x, y) = \frac{(1 - x^2 y^2)^2 + x^2 (1 - y^2)^2 + x^2 (1 - x^2)^2 y^2}{1 + x^2}$$

$$= \left(\frac{1 + x^2 - 2x^2 y^2}{1 + x^2} \right)^2 + \left(\frac{x(1 - x^2)y^2}{1 + x^2} \right)^2$$

$$+ \left(\frac{x(1 - x^2)y}{1 + x^2} \right)^2 + \left(\frac{x^2(1 - x^2)y}{1 + x^2} \right)^2.$$

2) Assume that $p(x,y) = \sum_{i=1}^{n} f_i(x,y)^2$ for some $n \in \mathbb{N}$ and polynomials $f_i(x,y) \in \mathbb{R}[x,y]$. Comparing terms of the same total degree on both sides we find $\deg f_i \leq 3$ for all i. Furthermore $p(0,y) = p(x,0) = 1$ implies that the polynomials $f_i(0,y)$ and $f_i(x,0)$ are constant. Altogether this gives

$$f_i(x,y) = a_i + xy\ell_i(x,y)$$

with $a_i \in \mathbb{R}$ and linear polynomials

$$\ell_i(x,y) = b_i + c_i x + d_i y \in \mathbb{R}[x,y].$$

But then the coefficient of the term $x^2 y^2$ in $\sum_1^n f_i(x,y)^2$ equals $\sum_1^n b_i^2$ which cannot be -3.

§3. The Subform Theorem

As in sections 1 and 2 we suppose char $K \neq 2$. Theorem 2.2 gives at least the following weak result for polynomials in several variables:

3.1 Proposition. (Substitution Principle) Let φ be an n-ary quadratic form over K, let $0 \neq p = p(t_1,\ldots,t_m) \in K[t_1,\ldots,t_m]$ be a polynomial and let $c_1,\ldots,c_m$ be arbitrary elements in K. If φ represents p over the rational function field $K(t_1,\ldots,t_m)$, then φ represents the element $p(c_1,\ldots,c_m)$ over K. (In the case $p(c_1,\ldots,c_m) = 0$ this may be the trivial representation of 0.)

PROOF by induction on m.
Theorem 2.2 implies that φ represents $p(t_1,\ldots,t_m)$ over $K(t_1,\ldots,t_{m-1})[t_m]$. Substituting c_m for t_m we see that φ represents $p(t_1,\ldots,t_{m-1},c_m)$ over the field $K(t_1,\ldots,t_{m-1})$. The assertion follows by induction.

Note. The substitution $t_m \to c_m$ could be impossible in a representation of p over the field $K(t_1,\ldots,t_m)$ since the denominators of the rational functions f_i in a representation $\varphi(f_1,\ldots,f_n) = p$ could vanish under $t_m \to c_m$. If, however, these denominators do not depend on the variable t_m they remain unchanged under the substitution.

3.2 Theorem. Let $d, a_1,\ldots,a_n \in K^{\bullet}$ and assume that $\varphi = \langle a_1,\ldots,a_n \rangle$ represents the polynomial $d + a_1 t^2$ over $K(t)$. Then either φ is isotropic over K or $\varphi' = \langle a_2,\ldots,a_n \rangle$ represents d over K.

PROOF. Suppose φ is anisotropic. By Theorem 2.2 we get

$$(1) \qquad a_1 f_1^2 + \ldots + a_n f_n^2 = d + a_1 t^2$$

where $f_i(t) \in K[t]$. Comparing terms of highest degree on both sides we conclude that all f_i must be linear in t, say $f_i(t) = b_i + c_i t$ $(i = 1, \ldots, n)$. Since char $K \neq 2$ at least one of the equations

$$b_1 + c_1 t = \pm t$$

is soluble in K, say with $t = c \in K$. Then (1) implies

$$(2) \qquad \sum_{i=2}^{n} a_i (b_i + c_i c)^2 = d,$$

i.e. φ' represents d over K.

3.3 Corollary. Let K be a field such that $\varphi = \underbrace{\langle 1, \ldots, 1 \rangle}_{n}$ is anisotropic over K, i.e. a nontrivial sum of n squares in K is never zero. (For example $K = \mathbf{Q}$ or $K = \mathbf{R}$.) Then $1 + t_1^2 + \ldots + t_n^2$ is not a sum of n squares in the rational function field $K(t_1, \ldots, t_n)$. Similarly, $t_1^2 + \ldots + t_n^2$ is not a sum of $n-1$ squares.

PROOF. Assume that φ represents $1 + t_1^2 + \ldots + t_n^2$. Apply 3.2 with $t = t_n$, $K(t_1, \ldots, t_{n-1})$ instead of K and $d = 1 + t_1^2 + \ldots + t_{n-1}^2$. Note that φ remains anisotropic over $K(t_1, \ldots, t_{n-1})$. It follows that $1 + t_1^2 + \ldots + t_{n-1}^2$ is represented by $\varphi' = \underbrace{\langle 1, \ldots, 1 \rangle}_{n-1}$ over $K(t_1, \ldots, t_{n-1})$. Continuing this process we see that $1 + t_1^2$ is a square in $K(t_1)$: contradiction.

The second statement is an immediate consequence of the first.

3.4 Theorem. (Subform Theorem) Let $\varphi \cong \langle a_1, \ldots, a_n \rangle$, $\psi \cong \langle b_1, \ldots, b_m \rangle$ be regular quadratic forms over K. Suppose φ anisotropic. Then the following statements are equivalent:

(1) ψ is isometric to a subform of φ, i.e.

$$\varphi \cong \psi \oplus \chi$$

for a suitable quadratic form χ over K. (Possibly $\chi = 0$ is the empty form of dimension 0.)

(2) $D_L(\psi) \subseteq D_L(\varphi)$ for every field $L \supseteq K$ (see Definition 1.10).

(3) φ represents the "generic value of ψ", i.e. φ represents

$$\psi(t_1, \ldots, t_m) = b_1 t_1^2 + \ldots + b_m t_m^2$$

over the rational function field $K(t_1, \ldots, t_m)$.

In particular all three statements imply $m \leq n$.

PROOF. The implications (1) $\Rightarrow$ (2) $\Rightarrow$ (3) are obvious. We prove (3) $\Rightarrow$ (1) by induction on $m = \dim \psi$, the case $m = 0$ being trivial. Suppose now $m > 0$. By 3.1 the form φ represents the element $b_1 \neq 0$ over K. (Put $t_1 = 1, t_2 = \ldots = t_m = 0$). By the note after Theorem 1.8 we can write $\varphi \cong \langle b_1 \rangle \oplus \varphi'$, where φ' is automatically anisotropic. Since φ represents $b_1 t_1^2 + (b_2 t_2^2 + \ldots + b_m t_m^2)$ over $K(t_2, \ldots, t_m)(t_1)$ we conclude from 3.2 with $d = b_2 t_2^2 + \ldots + b_m t_m^2$ that φ' represents $d = \psi'(t_2, \ldots, t_m)$ if we denote the form $\langle b_2, \ldots, b_m \rangle$ by ψ'. The induction hypothesis can now be applied to the pair φ', ψ'. This gives $\varphi' \cong \psi' \oplus \chi$ and

$$\varphi \cong \langle b_1 \rangle \oplus \varphi' \cong \langle b_1 \rangle \oplus \psi' \oplus \chi \cong \psi \oplus \chi.$$

Historical Note. For $\varphi = \langle \underbrace{1, \ldots, 1}_{n} \rangle$ Theorem 2.2, Theorem 3.2 and Corollary 3.3 are due to [Cassels 1964]. The influence of this paper on the development of the algebraic theory of quadratic forms was enormous. Together with the discovery of the multiplicative forms (see next chapter) it formed the basis for a thorough research on the Witt ring W and on the invariants s, p, u of a field (see Chapters 3, 7, 8). The generalization of the results of Cassels to an arbitrary quadratic form φ (as it is presented here) was immediate. It was first published in my paper [Pfister 1965$_2$].

§4. Appendix: The case char $K = 2$

For char $K = 2$ a lot of changes in the definitions, propositions and theorems have to be made but essentially most of the results of sections 1–3 carry over to this case. I shall indicate the necessary changes carefully but leave the details of some proofs to the reader. Filling in these details would be a good test for the right understanding of the previous sections!

Definitions 1.1 and 1.2 of a quadratic form and of equivalence were given without restriction on the characteristic. They thus remain in force. But contrary to the case char $K \neq 2$ the matrix A of the quadratic form

$$\varphi(x) = x'Ax = \sum_{i,j=1}^{n} a_{ij}x_ix_j = \sum_i a_{ii}x_i^2 + \sum_{i<j}(a_{ij} + a_{ji})x_ix_j$$

cannot be assumed to be symmetric. For char $K = 2$ φ remains unchanged iff A is replaced by $A + S$ where $S = (s_{ij})$ is a so-called *alternating* matrix, i.e. $s_{ii} = 0$, $s_{ij} = -s_{ji} = s_{ji}$ for $i \neq j$. This allows us for instance to take A as an upper triangular matrix, i.e. $a_{ij} = 0$ for $i > j$. Under an equivalence

$T \in GL_n(K)$ A goes to $T'AT$ and S goes to $T'ST$ which is again alternating. But if A is triangular then in general $T'AT$ is no longer triangular. Therefore the restriction to triangular matrices A is only of minor use.

The first essential change appears in the definition of the associated symmetric bilinear form b_φ of the quadratic form φ which is as follows:

4.1 Definition. For char $K = 2$ and $\varphi(x) = x'Ax$ put

$$b_\varphi(x, y) = \varphi(x + y) - \varphi(x) - \varphi(y) = x'(A + A')y.$$

Note that $A + A'$ does not change if A is replaced by $A + S$ with alternating S and that

$$b_\varphi(x, x) = 2\varphi(x) = 0 \quad \text{(identically in } x\text{)}.$$

This definition has the consequence that the study of *quadratic forms* $\varphi(x) = x'Ax$ (where A is an arbitrary $n \times n$ matrix) and of *symmetric bilinear forms*

$$b(x, y) = x'Ay$$

(where $A = A'$ is a symmetric matrix) falls apart for char $K = 2$. The theory of symmetric bilinear forms can be developed without even mentioning quadratic forms which is to a certain extent the point of view in the book [MH] whereas the theory of quadratic forms is intimately related to but richer than the theory of symmetric bilinear forms.

The geometric language of quadratic spaces and isometries between them can be used in characteristic 2 as well. Of course the factor $\frac{1}{2}$ in Definition 1.4 (2) has then to be cancelled. 1.5, 1.6 and 1.7 go over verbatim. Theorem 1.8 does not hold for char $K = 2$. It has to be replaced by Theorem 4.3 below. We first introduce the definition of the radical and of regularity:

4.2 Definition. Let (V, φ) be a quadratic space with associated symmetric bilinear form b_φ as in 4.1.

(1) For any subspace $U \subseteq V$ let

$$U^\perp = \{w \in V : b_\varphi(w, u) = 0 \text{ for all } u \in U\}$$

be its orthogonal space.

(2) The subspace rad $V = V^\perp$ is called the *radical* of (V, φ).

(3) (V, φ) is called *regular* if rad $V = 0$.

Note that for $\varphi(x) = x'Ax$ the form φ is regular if and only if $\det(A + A') \neq 0$. In particular, φ is never regular for char $K = 2$ if A is a diagonal matrix.

Regularity is a much stronger restriction here than for char $K \neq 2$. Instead of 1.8 we get

4.3 Theorem.

(1) Every quadratic space (V, φ) has an orthogonal decomposition

$$V = U \oplus \operatorname{rad} V$$

such that $(U, \varphi|_U)$ is regular.

(2) $\varphi|_{\operatorname{rad} V} \cong \langle d_1 \rangle \oplus \ldots \oplus \langle d_r \rangle$ with $r = \dim(\operatorname{rad} V)$ and $d_i \in K$ is an orthogonal sum of 1-dimensional spaces.

(3) $(U, \varphi|_U)$ is an orthogonal sum of 2-dimensional regular quadratic spaces (U_i, φ_i), $i = 1, \ldots, s$.

PROOF. 1) For *every* complement U of $\operatorname{rad} V$ the form $\varphi|_U$ is regular since a vector $w \in U^\perp$ is actually in $(U \oplus \operatorname{rad} V)^\perp = V^\perp = \operatorname{rad} V$, hence $w \in U \cap U^\perp$ implies $w \in U \cap \operatorname{rad} V = \{0\}$.
2) Clearly every basis of $\operatorname{rad} V$ is an orthogonal basis with respect to b_φ since $b_\varphi|_{\operatorname{rad} V \times \operatorname{rad} V}$ is trivial.
3) The case $U = 0$ is trivial. Otherwise take an arbitrary $0 \neq u_1 \in U$ and choose $u_2 \in U$ such that $b_\varphi(u_1, u_2) = b \neq 0$. Put $U_1 = K u_1 + K u_2$. Since $b_\varphi(u_1, u_1) = 2\varphi(u_1) = 0$ the vector u_2 is linearly independent of u_1, i.e. $\dim U_1 = 2$. Let $\varphi(u_1) = a$, $\varphi(u_2) = c$. Then we find

$$\begin{aligned} \varphi(x_1 u_1 + x_2 u_2) &= \varphi(x_1 u_1) + b_\varphi(x_1 u_1, x_2 u_2) + \varphi(x_2 u_2) \\ &= ax_1^2 + bx_1 x_2 + cx_2^2. \end{aligned}$$

This says that $\varphi_1 = \varphi|_{U_1}$ has matrix $A = \begin{pmatrix} a & b \\ 0 & c \end{pmatrix}$ and implies that φ_1 is regular since $A + A' = \begin{pmatrix} 0 & b \\ b & 0 \end{pmatrix}$ has non-vanishing determinant.

Finally $U = U_1 \oplus \tilde{U}$ with $\tilde{U} = U \cap U_1^\perp$ is an orthogonal decomposition of U and $\varphi|_{\tilde{U}}$ must be regular. Replace now U by $\tilde{U}$ and continue with splitting off regular 2-dimensional subspaces. This finishes the proof.

Notation. On the way we have found the typical matrix $A = \begin{pmatrix} a & b \\ 0 & c \end{pmatrix}$ with $a, b, c \in K$, $b \neq 0$, of a regular 2-dimensional space (V, φ). This space will be denoted by $[a, b, c]$. Scaling the second basis vector with b^{-1} we can always suppose that $b = 1$.

In the decomposition $V = U \oplus \operatorname{rad} V$ from 4.3 the subspace $(\operatorname{rad} V, \varphi|_{\operatorname{rad} V})$ of (V, φ) is clearly unique. But the "regular part" $(U, \varphi|_U)$ of (V, φ) is not unique, not even up to isometry. This new phenomenon is shown by

4.4 Example. For a field K of characteristic 2 we have

$$[1, 1, 1] \oplus \langle 1 \rangle \cong [0, 1, 0] \oplus \langle 1 \rangle$$

but $[1, 1, 1] \cong [0, 1, 0]$ holds if and only if the quadratic equation $x^2 + x + 1 = 0$ has a solution in K.

Taking over verbatim Definition 1.10 to the case char $K = 2$ we also find

4.5 Proposition. Up to isometry there is just one regular isotropic quadratic space of dimension 2, namely the hyperbolic plane H with the form $\varphi = [0, 1, 0]$, $\varphi(x) = x_1 x_2$.

The simple proofs of 4.4 and 4.5 are left to the reader.

Similarly, Proposition 1.12 also holds for char $K = 2$. We only have to replace the form $\langle 1, -1 \rangle$ by $[0,1,0]$.

The subspace $(\operatorname{rad} V, \varphi|_{\operatorname{rad} V})$ of (V, φ) in Theorem 4.3 is often called the *quasilinear part* of (V, φ). This is explained by the rules

$$\varphi(ax) = a^2 \varphi(x), \quad \varphi(x + y) = \varphi(x) + \varphi(y)$$

for $a \in K$ and $x, y \in \operatorname{rad} V$.

For char $K = 2$ the subset $K_0 = K^2$ of all squares of K is a subfield of K, since $a^2 b^2 = (ab)^2$ and $a^2 + b^2 = (a + b)^2$. Therefore K may be considered as a vector-space over K_0 (which can be of finite or infinite dimension). If now $\varphi|_{\operatorname{rad} V} = \langle d_1, \ldots, d_r \rangle$ is isotropic over K (which is automatically true for $r > \dim_{K_0} K$), say

$$d_1 a_1^2 + \ldots + d_r a_r^2 = 0$$

with $a_i \in K$, $a_r \neq 0$, then

$$\sum_1^r d_i x_i^2 = \sum_1^{r-1} d_i \left(x_i + \frac{a_i}{a_r} x_r \right)^2 + 0 \cdot x_r^2,$$

i.e. $\langle d_1, \ldots, d_r \rangle \cong \langle d_1, \ldots, d_{r-1}, 0 \rangle$.

This leads to the next definition and proposition.

4.6 Definition. An n-ary quadratic form φ over K or its space (V, φ), is called *non-defective* if there is no linear transformation $T \in GL_n(K)$ such that $\varphi(Tx)$ is independent of x_n.

4.7 Proposition. (V, φ) is non-defective iff $\varphi|_{\operatorname{rad} V}$ is anisotropic.

The proof is left to the reader.

Note. The property "non-defective" is not invariant under field extensions L/K.

From now on we may always assume that all our spaces (V, φ) are non-defective. However, we cannot always assume that (V, φ) is regular as for char $K \neq 2$. Obviously every anisotropic space is non-defective.

We now have to adjust sections 2 and 3 to the case char $K = 2$. As regards 2.1–3 this is easy since we did not make use of the fact that φ can be diagonalized for char $K \neq 2$: The proof of 2.1 did not depend on char K at all. In the proof of 2.2 we may now suppose that φ is non-defective. If such a φ is isotropic over K then for every isotropic $0 \neq u_1 \in V$ we find $u_2 \in V$ such that $b_\varphi(u_1, u_2) \neq 0$ (since otherwise $u_1 \in \operatorname{rad} V = 0$). But then the proof of 4.3 shows that (V, φ) contains the hyperbolic plane H as an orthogonal summand. This implies

$$\varphi(x) = x_1 x_2 + \psi(x_3, \ldots, x_n)$$

as in part 2) of the proof of Theorem 2.2. From this we immediately get a representation of p over $K[t]$, namely $x_1 = p$, $x_2 = 1$, $x_3 = \ldots = x_n = 0$. Finally, if φ is anisotropic, part 3) of the proof of Theorem 2.2 carries over nearly verbatim. The only change occurs in formulas (4)–(6) where $\mu = 2b_\psi(f, g)$ has to be replaced by $\mu = b_\psi(f, g)$. Altogether we see that 2.1, 2.2 and 2.3 hold for char $K = 2$ as well.

Clearly the same is true for Proposition 3.1.

Theorem 3.2 has to be replaced by the following.

4.8 Theorem. Let $a, b, c, d \in K$ with $bd \neq 0$ and assume that the K-anisotropic quadratic form φ represents

$$at_1^2 + bt_1 t_2 + ct_2^2 + d$$

over the field $K(t_1, t_2)$. Then $\varphi \cong [a, b, c] \oplus \chi$ (over K) such that χ represents d over K.

SKETCH OF PROOF.

(1) Show that φ represents $at_1^2 + bt_1 + c$ over $K(t_1)$ and d over K. (Use 3.1.)

(2) By 2.2 φ represents $at_1^2 + bt_1 + c$ over $K[t_1]$, say $\varphi(f_1, \ldots, f_n) = at_1^2 + bt_1 + c$, $f_i \in K[t_1]$. Since φ is anisotropic deduce that the f_i are linear. Compare coefficients, find vectors u, v over K such that $\varphi(u) = a$, $b_\varphi(u, v) = b$, $\varphi(v) = c$.

(3) Repeat step (2) for a representation of

$$at_1^2 + (bt_2)t_1 + (ct_2^2 + d) \quad \text{over } K(t_2)[t_1]$$

and find vectors u', w' over $K(t_2)$ with

$$\varphi(u') = a, \quad b_\varphi(u', w') = bt_2, \quad \varphi(w') = ct_2^2 + d.$$

If $u' = u$ put $w = w'$. Otherwise $u - u' = u + u' \neq 0$, hence $\varphi(u + u') = \varphi(u) + \varphi(u') + b_\varphi(u, u') = 2a + b_\varphi(u, u') \neq 0$ since φ is anisotropic, i.e. $b_\varphi(u, u') \neq 0$. Consider the linear transformation T of $K(t_2)^n$ which interchanges u and u' and is the identity on the orthogonal complement of the (regular) 2-dimensional subspace generated by u and u'. Then T is a self-isometry of $\varphi \otimes K(t_2)$ since $\varphi(u) = \varphi(u') = a$. Put $Tw' = w$. Then in any case

$$(*) \qquad \varphi(u) = a, \ b_\varphi(u, w) = bt_2, \ \varphi(w) = ct_2^2 + d.$$

(4) Show by using 2.2 again that w may be taken to have components in $K[t_2]$. Then

$$w = w_1 t_2 + w_0$$

with vectors w_0, w_1 over K since φ is anisotropic.

(5) Let U be the subspace of $V = K^n$ which is spanned by u, w_1. Deduce from $(*)$ that $w_0 \in U^\perp$ and

$$\varphi|_U \cong [a, b, c], \ \varphi(w_0) = d.$$

This implies our assertion with $\chi = \varphi|_{U^\perp}$.

Finally, Theorem 3.4 has to be replaced by

4.9 Theorem Let (V, φ) and (W, ψ) be non-defective quadratic spaces over K, $\dim V = n$, $\dim W = m$. Suppose φ anisotropic. Then the following statements are equivalent:

(1) ψ is isometric to a subform of φ, i.e. there exists a subspace U of V such that $\varphi|_U \cong \psi$.
   (Note: U need not be an orthogonal summand of V !)

(2) $D_L(\psi) \subseteq D_L(\varphi)$ for every field $L \supseteq K$.

(3) φ represents $\psi(t_1, \ldots, t_m)$ over the rational function field $K(t_1, \ldots, t_m)$.

SKETCH OF PROOF. As for 3.4 we only have to show (3) $\Rightarrow$ (1). By Theorem 4.3 we have

$$(W, \psi) = W_1 \oplus \ldots \oplus W_s \oplus \operatorname{rad} W$$

with

$$\psi|_{W_i} \cong [a_i, b_i, c_i] \quad (i = 1, \ldots, s),$$
$$\psi|_{\operatorname{rad} W} \cong \langle d_1 \rangle \oplus \ldots \oplus \langle d_r \rangle$$
$$\text{and} \quad 2s + r = m.$$

(1) Apply 4.8 s times and get an orthogonal decomposition

$$(V, \varphi) = U_1 \oplus \ldots \oplus U_s \oplus V_0$$

such that $\varphi|_{U_i} \cong [a_i, b_i, c_i]$ and that $\varphi|_{V_0} = \varphi_0$ represents

$$d_1 t_1^2 + \ldots + d_r t_r^2 \quad \text{over } K(t_1, \ldots, t_r).$$

(2) Show by induction on r that V_0 contains a subspace V_1 with $\varphi|_{V_1} \cong \langle d_1, \ldots, d_r \rangle$. This is trivial for $r = 0$ or $r = 1$. For $r \geq 2$ put $d' = d_2 t_2^2 + \ldots + d_r t_r^2$. φ_0 represents $d_1 t_1^2 + d'$ over $K(t_2, \ldots, t_r)(t_1)$ hence over $K(t_2, \ldots, t_r)[t_1]$. This produces vectors v_1' and v' over $K(t_2, \ldots, t_r)$ such that $\varphi_0(v_1') = d_1$, $\varphi_0(v') = d'$ and $b_{\varphi_0}(v_1', v') = 0$. On the other hand we also know that φ_0 represents d_1 over K, i.e. there is a vector v_1 over K with $\varphi_0(v_1) = d_1$. As in step (3) of the last proof either we have $v_1' = v_1$ or v_1, v_1' span a regular 2-dimensional subspace and there is a "switch" $v_1' \to v_1$, $v' \to v$ over the field $K(t_2, \ldots, t_r)$. In any case we get vectors v_1 over K and v over $K(t_2, \ldots, t_r)$ such that

$$\varphi_0(v_1) = d, \ \varphi_0(v) = d', \ b_\varphi(v_1, v) = 0.$$

By the induction hypothesis the equation $\varphi_0(v) = d'$ leads to a subspace V_2 of V_0 (over K) with

$$\varphi_0|_{V_2} \cong \langle d_2, \ldots, d_r \rangle.$$

For any vector $v_2 \in V_2$ we have $\varphi_0(v_2) = d_2 a_2^2 + \ldots + d_r a_r^2$ with $a_2, \ldots, a_r \in K$. Since $\psi|_{\mathrm{rad}\, w}$ is anisotropic this implies $\varphi_0(v_2) \neq d_1 = \varphi_0(v_1)$ for all $v_2 \in V_2$, i.e. $v_1 \notin V_2$. Put $V_1 = K v_1 + V_2$, $U = U_1 \oplus \ldots \oplus U_s \oplus V_1$. Then $\varphi|_U \cong \psi$ and the proof is finished.

Historical Note. The theory of quadratic forms over a field of characteristic 2 was developed by C. Arf [1941]. In particular Theorem 4.3 is due to him. Theorems 3.2 (for char $K = 2$), 4.8 and 4.9 were first proved in a diploma paper of M. Amer (Göttingen 1970).

Chapter 2

Multiplicative Quadratic Forms

§1. The Theorem of Witt

As stated in the preface we shall not develop the full algebraic theory of quadratic forms here since this has been done in a very satisfactory way in other books, in particular [L] and [S]. But Witt's theorem and the definition and existence of the Witt ring should not be missing in any book on quadratic forms. Without them we could not even understand *why* multiplicative forms are important for the theory.

As in Chapter 1 we will asume char $K \neq 2$ for our treatment but indicate the differences which occur for char $K = 2$ in an appendix.

1.1 Theorem (Witt's Cancellation Theorem)
Let $\varphi, \varphi_1, \varphi_2$ be quadratic forms over K such that

$$\varphi \oplus \varphi_1 \cong \varphi \oplus \varphi_2 \quad (\text{over } K).$$

Then $\varphi_1 \cong \varphi_2$ (over K).

PROOF. 1) By Theorem 1.8 of Chapter 1 all three forms may be supposed to be diagonal since neither the assumption nor the assertion of the theorem is changed if we replace φ or φ_1 or φ_2 by an equivalent form. Let, say, $\varphi(x) = x'Ax$ with diagonal matrix

$$A = \begin{pmatrix} a_1 & & \\ & \ddots & \\ & & a_m \end{pmatrix}$$

where $m = \dim \varphi$. Let $r = r(\varphi)$ denote the rank of A. Then the number of zeros among the entries $a_1, \ldots, a_m$ equals $m - r$. Since an equivalence relation does not change dimension and rank we conclude from the assumption $\varphi \oplus \varphi_1 \cong \varphi \oplus \varphi_2$ that

$$\dim \varphi_1 = \dim \varphi_2 = n \text{ (say)}, \quad r(\varphi_1) = r(\varphi_2).$$

This allows us to assume that φ_1 and φ_2 are regular. (Otherwise take away the 0-entries from φ_1 and φ_2 and "add" them to φ.) In addition we may assume by a trivial induction on $\dim \varphi$ that $\dim \varphi = 1$, i.e. $\varphi \cong \langle a \rangle$ with $a \in K$.

2) After these steps of reduction the hypothesis gives an $(n+1) \times (n+1)$-matrix T with entries from K such that

$$(\varphi \oplus \varphi_1)(Tx) = (\varphi \oplus \varphi_2)(x).$$

Let $x = \begin{pmatrix} x_0 \\ y \end{pmatrix}$ where x_0 is a scalar and y is a column vector of length n. Similarly

$$T = \begin{pmatrix} t & u' \\ v & U \end{pmatrix}$$

with $t \in K$, $u, v \in K^n$, $U \in M_n(K)$. We get

$$a(tx_0 + u'y)^2 + \varphi_1(vx_0 + Uy) = ax_0^2 + \varphi_2(y).$$

Since char $K \neq 2$ at least one of the two equations

$$\pm x_0 = tx_0 + u'y$$

has a solution $x_0 = \frac{u'}{\pm 1 - t} \cdot y \in K$.
   Put $w = \frac{u}{\pm 1 - t} \in K^n$. Then

$$\varphi_1((vw' + U)y) = \varphi_2(y).$$

Since φ_1 and φ_2 are regular the $n \times n$-matrix $vw' + U$ must also be regular. Hence $\varphi_1 \cong \varphi_2$. Note the matrix equation $v(w'y) = (vw')y$ for the column vectors v, w, y !

Before we come to the definition of the Witt ring of K we introduce some more notation. From now on all quadratic forms are supposed to be regular and diagonal since we are only interested in operations and results for the *set of equivalence classes of quadratic forms over K*.

1.2 Notation.

(1) $G(K) = K^\bullet / K^{\bullet 2}$ is the *square class group of K*.

(2) For $\varphi \cong \langle a_1, \ldots, a_m \rangle$ with $a_i \in K^\bullet$
     $\det \varphi = (\prod_1^m a_i) K^{\bullet 2} \in G(K)$ is the *determinant of φ*
     and
     $d(\varphi) = (-1)^{\frac{m(m-1)}{2}} \det \varphi \in G(K)$ is the *discriminant of φ*.

(3) For φ as in (2) and $\psi \cong \langle b_1, \ldots, b_n \rangle$
     $\varphi \oplus \psi \cong \langle a_1, \ldots, a_m, b_1, \ldots, b_n \rangle$ is the (orthogonal) *sum of φ and ψ* (see also 1.7 of Chapter 1)
     and
     $\varphi \otimes \psi \cong \langle \ldots, a_i b_j, \ldots \rangle_{\substack{i=1,\ldots,m \\ j=1,\ldots,n}}$ is the (tensor) *product of φ and ψ*.

(4) For $\varphi \cong \langle a_1, \ldots, a_m \rangle$ and $a \in K^\bullet$
     $a\varphi \cong \langle aa_1, \ldots, aa_m \rangle$ is the *scaling of φ by a*.

(5) For $r \in \mathbb{N}_0$ and any φ

$$r \times \varphi \cong \underbrace{\varphi \oplus \ldots \oplus \varphi}_{r \text{ times}}$$

is the *r-fold addition of φ to itself*.

(6) $0 \times \varphi = 0$ is the empty form of dimension 0.

Notes. Since $\langle a \rangle \cong \langle ac^2 \rangle$ for $a, c \in K^{\bullet}$ the diagonal entries of $\varphi \cong \langle a_1, \ldots, a_m \rangle$ matter only up to square factors. We might take $a_i \in G(K)$ instead of $a_i \in K^{\bullet}$. If φ is replaced by an equivalent form φ_1 the symmetric matrix A of φ is replaced by $A_1 = T'AT$ with $T \in GL_n(K)$. Then $\det A_1 = \det A \cdot (\det T)^2$. Therefore the determinant and the discriminant of (the equivalence class of) φ are well-defined as elements of $G(K)$. Up to equivalence the order of the entries in the definition of $\varphi \oplus \psi$ and $\varphi \otimes \psi$ plays no role. We have $\dim(\varphi \oplus \psi) = \dim \varphi + \dim \psi$ and $\dim(\varphi \otimes \psi) = \dim \varphi \cdot \dim \psi$.

A very simple but useful conclusion of the invariance of the determinant under equivalence is the following.

1.3 Lemma. If $\langle a, b \rangle$ represents $c \in K^{\bullet}$ then

$$\langle a, b \rangle \cong \langle c, abc \rangle.$$

PROOF. By the note after Theorem 1.8 of Chapter 1 we have $\langle a, b \rangle \cong \langle c, e \rangle$ for some $e \in K^{\bullet}$. Comparing determinants yields $abce \in K^{\bullet 2}$ hence $\langle e \rangle \cong \langle abc \rangle$.

The next result is also immediate.

1.4 Proposition. Let φ be a (regular) quadratic form over K and let $c \in K^{\bullet}$. Then we have

$$\varphi \text{ represents } c \text{ (over } K) \Longleftrightarrow \varphi \oplus \langle -c \rangle \text{ isotropic (over } K).$$

PROOF. 1) If φ represents c then $\varphi(v) = c$ for some vector $v \in K^n$ where $n = \dim \varphi$. Therefore

$$(\varphi \oplus \langle -c \rangle) \binom{v}{1} = \varphi(v) - c \cdot 1^2 = c - c = 0,$$

i.e. $\varphi \oplus \langle -c \rangle$ is isotropic.

2) Let $\varphi \oplus \langle -c \rangle$ be isotropic. If φ is itself isotropic then we see from Propositions 1.11 and 1.12 of Chapter 1 that φ is universal, i.e. $c \in D_K(\varphi)$. If φ is anisotropic then the isotropy of $\varphi \oplus \langle -c \rangle$ gives an equation $\varphi(v) - cv_{n+1}^2 = 0$ in K with $v_{n+1} \in K^{\bullet}$. Then $\varphi\left(\frac{v}{v_{n+1}}\right) = c$.

Combining Witt's Cancellation Theorem with Proposition 1.12 of Chapter 1 we get

1.5 Theorem. (Witt Decomposition) Every (regular) quadratic form φ over K has an orthogonal decomposition

$$\varphi \cong i \times \langle 1, -1 \rangle \oplus \varphi_0$$

with $i \in \mathsf{N}_0$, φ_0 anisotropic over K. The *Witt index* i is uniquely determined by φ, φ_0 is unique up to equivalence.

PROOF. The existence of the decomposition comes from a repeated application of Proposition 1.12 of Chapter 1, the uniqueness is a direct consequence of the Cancellation Theorem 1.1.

Notation. φ_0 is called the *anisotropic part* or *kernel* of φ.

1.6 Note. By 1.5 the full classification of (finite-dimensional) quadratic forms over K is reduced to the following two problems:

(A) A method which decides whether a given form φ is isotropic or not.

(B) The classification of the anisotropic forms.

For special classes of fields, notably the fields which turn up in number theory, both problems can be fully solved. But for general K only partial results are known.

Concerning (A) they are of the kind: There is a numerical invariant $u = u(K)$ of the field K which is finite for many interesting classes of fields and such that every form φ with $\dim \varphi > u$ is isotropic over K (but the isotropy problem remains open for forms with $\dim \varphi \le u$). This u-invariant will be treated quite extensively in Chapters 8 and 9.

Concerning (B) we find that the anisotropic forms "make up" a commutative associative ring $W(K)$. Therefore (B) is considered to be equivalent to the computation of $W(K)$. Here quite a bit about the *structure* of $W(K)$ is known but a full description of $W(K)$ is available only for special classes of fields. We shall derive some easy properties of $W(K)$ and give some examples but refer to the "standard books" for a fuller treatment.

The Decomposition Theorem 1.5 allows a coarser equivalence relation for quadratic forms than ordinary equivalence, the so-called *similarity* or *Witt equivalence*:

1.7 Definition. Two (regular) quadratic forms φ, ψ over K which have anisotropic parts φ_0, ψ_0 are called *similar* if $\varphi_0 \cong \psi_0$. Similarity is denoted by $\varphi \sim \psi$.

In other words: φ and ψ are similar iff there exist integers $r, s \in \mathsf{N}_0$ such that

$$\varphi \cong r \times \langle 1, -1 \rangle \oplus \varphi_0, \quad \psi \cong s \times \langle 1, -1 \rangle \oplus \varphi_0.$$

Clearly $\sim$ is an equivalence relation.

1.8 Notation. The set of similarity classes $\tilde{\varphi}$ of regular quadratic forms φ of any finite dimension $n \geq 0$ over K is denoted by $W(K)$. Here the empty form $\varphi = 0$ of dimension 0 is counted as being regular and anisotropic.

1.9 Theorem. [Witt 1937]
The operations $\oplus, \otimes$ can be naturally defined on $W(K)$. Using $\oplus$ as addition and $\otimes$ as multiplication the set $W(K)$ becomes a commutative associative ring with zero element $0 = \tilde{0}$ and unit element $1 = \langle \tilde{1} \rangle$. It is called the *Witt ring* of K. If multiplication is ignored it is called the *Witt group* of K.

PROOF. 1) For $\tilde{\varphi}, \tilde{\psi} \in W(K)$ where φ, ψ are regular quadratic forms over K define

$$\tilde{\varphi} \oplus \tilde{\psi} = \widetilde{\varphi \oplus \psi}, \quad \tilde{\varphi} \otimes \tilde{\psi} = \widetilde{\varphi \otimes \psi}.$$

Then $\oplus$ and $\otimes$ are well-defined on $W(K)$. This is clear for $\oplus$. For $\otimes$ we have to use the fact that

$$\langle 1, -1 \rangle \otimes \langle a \rangle \cong \langle a, -a \rangle \cong \langle 1, -1 \rangle \quad \text{for every } a \in K^\bullet$$

which implies $\langle 1, -1 \rangle \otimes \psi \cong (\dim \psi) \times \langle 1, -1 \rangle$ for every form ψ. It also shows that 0 is a zero element and 1 is a unit element.

2) The commutative, associative and distributive laws for $\oplus$ and $\otimes$ on $W(K)$ are clear (compare 1.2). By 1.1 $W(K)$ satisfies the cancellation law, it is therefore a monoid.

3) Every element $\tilde{\varphi} = \langle a_1, \ldots, a_n \rangle \in W(K)$ has an additive inverse, namely $\widetilde{\varphi'} = \langle -a_1, \ldots, -a_n \rangle$. Without danger of confusion we also write

$$\widetilde{\varphi'} = \ominus \tilde{\varphi} \quad \text{and} \quad \tilde{\psi} \ominus \tilde{\varphi} = \tilde{\psi} \oplus \widetilde{\varphi'}.$$

This proves the theorem.

1.10 Examples.

(1) Let K be a quadratically closed field with char $K \neq 2$, e.g. $K = \mathbf{C}$. Then the only anisotropic quadratic forms over K are 0 and $\langle 1 \rangle$. Hence $W(K) = \mathbf{Z}/2\mathbf{Z}$.

(2) Let $K = \mathbf{R}$. Then the only anisotropic quadratic forms over K are 0, $n \times \langle 1 \rangle$ and $n \times \langle -1 \rangle$ where $n \in \mathbf{N}$. Hence $W(K) = \mathbf{Z}$.

(3) Let $K = \mathbf{F}_p$ be the finite field of p elements where p is an odd prime number.

   a) $K^\bullet$ has exactly two square classes represented by 1 and any non-square $\varepsilon \in K^\bullet$.

  b) ε is a sum of two squares in K.

  c) Every binary (i.e. 2-dimensional) quadratic form over K is universal.
     Hence $|W(K)| = 4$.

  d) As a ring $W(K) \cong \mathbf{Z}/2\mathbf{Z} \oplus \mathbf{Z}/2\mathbf{Z}$ for $p \equiv 1 \mod 4$
     and $\qquad W(K) \cong \mathbf{Z}/4\mathbf{Z} \qquad$ for $p \equiv 3 \mod 4$

PROOF.

(1) Since every $a \in K^\bullet$ is a square we have just one 1-dimensional form,
    namely $1 \cong \langle 1 \rangle$. Every 2-dimensional form $\langle a, b \rangle$ is isotropic since $b = -ac^2$ for some $c \in K^\bullet$. Therefore the only anisotropic forms are 0 and
    1. By definition of $W(K)$ every element $w \in W(K)$ is represented by an
    anisotropic form φ over K (which is unique up to equivalence). Hence
    $|W| = 2$ and $W = \mathbf{Z}/2\mathbf{Z}$ as a ring.

(2) $\mathbf{R}^\bullet$ has two square classes represented by 1 and -1. Therefore every n-
    dimensional $\varphi = \langle a_1, \ldots, a_n \rangle$ has the shape $\varphi = r \times \langle 1 \rangle \oplus s \times \langle -1 \rangle$,
    where $r, s \in \mathbf{N}_0$, $r + s = n$. Clearly φ is isotropic if $r > 0$ and $s > 0$. If,
    however, $s = 0$ then $\varphi = n \times \langle 1 \rangle = \underbrace{\langle 1, \ldots, 1 \rangle}_{n}$ is anisotropic since every
    nontrivial sum of n squares in $\mathbf{R}$ is positive and therefore different from 0
    (for $n > 0$). A similar argument works for $r = 0$, $s = n$. This shows that
    $W(K) = \mathbf{Z}$ as sets, and it is obvious that the ring structures on $W(K)$
    and $\mathbf{Z}$ coincide.

(3)   a) As is well-known from elementary number theory there are $\frac{p-1}{2}$
         quadratic residues and as many non-residues modulo p. This shows
         $K^\bullet = K^{\bullet 2} \cup \varepsilon K^{\bullet 2}$ for $K = \mathbf{F}_p = \mathbf{Z}/p\mathbf{Z}$. Alternatively the group-
         homomorphism of $K^\bullet$ into itself which sends x to x^2 has kernel ± 1.
         Therefore its image $K^{\bullet 2}$ must contain $\frac{p-1}{2}$ elements.

      b) The subsets K^2 and $\varepsilon - K^2$ of K both have $\frac{p+1}{2}$ elements. Therefore
         their intersection is non-empty, i.e. there exist $a, b \in K$ such that
         $a^2 = \varepsilon - b^2$.

      c) Every anisotropic binary form over $\mathbf{F}_p$ is a scalar multiple of $\varphi_0 = \langle 1, -\varepsilon \rangle$. It suffices to show that φ_0 is universal, i.e. represents 1 and
         ε. This follows from $1 = 1 \cdot 1^2 - \varepsilon \cdot 0^2$, $\varepsilon = (\frac{\varepsilon}{b})^2 - \varepsilon(\frac{a}{b})^2$. Note that
         $ab \neq 0$ since ε is not a square. From 1.4 we see that every form of
         dimension ≥ 3 is isotropic. The elements of $W(K)$ are then given
         by the anisotropic forms

$$0, \ 1 = \langle 1 \rangle, \ \langle \varepsilon \rangle, \ \varphi_0.$$

      d) The Legendre symbol $(\frac{-1}{p})$ is ± 1 according as $p \equiv \pm 1 \mod 4$. In
         the first case $\langle 1, 1 \rangle \sim \langle 1, -1 \rangle \sim 0$ which means that $1 = \langle 1 \rangle$ (and

therefore every element $\neq 0$ of W) has order 2 in the group W.
In the second case $2 \times \langle 1 \rangle = \langle 1, 1 \rangle \not\sim 0$ but of course $4 \times \langle 1 \rangle \sim 0$
since $|W| = 4$.

This finishes the proof.

Note. Example (3) can be generalized to odd prime powers $q = p^m$ instead
of primes p.

§2. The Multiplicative Forms

As before let K be an arbitrary field of characteristic different from 2. For a
regular quadratic form φ of dimension $n \geq 1$ over K we consider two independent indeterminate column vectors x, y of length n over K. The corresponding
row vectors are then given as $x' = (x_1, \ldots, x_n)$, $y' = (y_1, \ldots, y_n)$. The rational
function fields which are determined by x, resp. x and y, are denoted as follows:
$K(x) = K(x_1, \ldots, x_n)$,
$K(x, y) = K(x_1, \ldots, x_n, y_1, \ldots, y_n)$.

2.1 Definition.

a) φ is called *multiplicative* (over K) if there exists a vector $z' = (z_1, \ldots, z_n)$
   with entries $z_i \in K(x, y)$ such that the following equation holds in
   $K(x, y)$:
   (1) $\qquad\qquad \varphi(x) \cdot \varphi(y) = \varphi(z)$.

b) φ is called *strictly multiplicative* (over K) if z can be chosen to depend
   linearly on y, i.e. if there exists a matrix $T_x \in M_n(K(x))$ such that (1)
   holds with $z = T_x y$.
   In other words, we have $\varphi(x)\varphi(y) = \varphi(T_x y)$ identically in x and y, or
   (since T_x must be regular)

   (2) $\qquad\qquad \varphi(x)\varphi \cong \varphi \quad$ over $K(x)$.

   For $\varphi(x) = x'Ax$ with symmetric matrix $A = A' \in M_n(K)$ condition (2)
   may be expressed by
   (2') $\qquad\qquad \varphi(x)A = T_x' A T_x \quad$ in $M_n(K(x))$.

The main result is the following theorem which implies in particular that
strictly multiplicative quadratic forms exist in all 2-power dimensions $n = 2^k$.

2.2 Theorem. For arbitrary $k \geq 0$ and $a_1, \ldots, a_k \in K^\bullet$ the 2^k-dimensional
form
$$\varphi = \langle 1, a_1 \rangle \otimes \ldots \otimes \langle 1, a_k \rangle$$

is strictly multiplicative over K.

FIRST PROOF (using matrices). We use induction on k. The case $k = 0$ is trivial: $\varphi = \langle 1 \rangle$, $\varphi(x) = x^2$, $A = (1)$. Here the matrix $T_x = (x) \in M_1(K(x))$ solves equation (2').

For the inductive step $k \to k+1$ we take φ as above and put $\psi = \varphi \otimes \langle 1, a \rangle = \varphi \oplus a\varphi$ with $a = a_{k+1} \in K^\bullet$. Let $n = 2^k$ and let A be the diagonal matrix for φ. By the induction hypothesis we have

$$\varphi(x) = x'Ax \quad \text{and} \quad \varphi(x)A = T_x'AT_x.$$

The diagonal matrix B of ψ is given by

$$B = \begin{pmatrix} A & 0 \\ 0 & aA \end{pmatrix} \in M_{2n}(K).$$

The "generic" value of ψ is given by

$$\psi(x, y) = \begin{pmatrix} x \\ y \end{pmatrix}' B \begin{pmatrix} x \\ y \end{pmatrix} = x'Ax + y'aAy = \varphi(x) + a\varphi(y)$$

where x, y are independent indeterminate column vectors of length n.

We have to show that there exists a matrix $T_{x,y} \in M_{2n}(K(x, y))$ such that
(*) $\psi(x, y)B = T_{x,y}'BT_{x,y}.$
Put

$$T_{x,y} := \begin{pmatrix} T_x & aT_y \\ -T_y & U \end{pmatrix} \quad \text{with } U := \varphi(y)^{-1}\varphi(x)T_yT_x^{-1}T_y.$$

Control of (*)
We have

$$BT_{x,y} = \begin{pmatrix} AT_x & aAT_y \\ -aAT_y & aAU \end{pmatrix}$$

and

$$T_{x,y}'BT_{x,y} = \begin{pmatrix} T_x'AT_x - T_y'(-aAT_y) & T_x'(aAT_y) - T_y'(aAU) \\ aT_y'AT_x + U'(-aAT_y) & aT_y'(aAT_y) + U'(aAU) \end{pmatrix}.$$

The four entries of this matrix are as follows:

$$
\begin{aligned}
1^{st} \text{ entry} &= \varphi(x)A + a\varphi(y)A = \psi(x, y)A; \\
2^{nd} \text{ entry} &= aT_x'AT_y - a\varphi(x)\varphi(y)^{-1}(T_y'AT_y)T_x^{-1}T_y \\
&= aT_x'AT_y - a(\varphi(x)A)T_x^{-1}T_y \\
&= aT_x'AT_y - a(T_x'AT_x)T_x^{-1}T_y = 0; \\
3^{rd} \text{ entry} &= \text{transpose of } 2^{nd} \text{ entry} = 0; \\
4^{th} \text{ entry} &= a^2\varphi(y)A + a\varphi(y)^{-2}\varphi(x)^2T_y'T_x'^{-1}(T_y'AT_y)T_x^{-1}T_y.
\end{aligned}
$$

Here the second summand simplifies to

$$a\varphi(y)^{-1}\varphi(x)^2 T_y'(T_x'^{-1}AT_x^{-1})T_y = a\varphi(y)^{-1}\varphi(x)T_y'AT_y = a\varphi(x)A.$$

Thus the fourth entry equals

$$a^2\varphi(y)A + a\varphi(x)A = \psi(x,y)aA.$$

This proves (*).

Notes.

(1) Since $\varphi(x)T_x^{-1} = A^{-1}T_x'A$ the matrix U can be written in the form $U = \varphi(y)^{-1}\, T_y\, A^{-1}\, T_x'\, AT_y$. This shows that U and $T_{x,y}$ depend "integrally" on T_x.

(2) On the other hand the factor $\varphi(y)^{-1}$ in U cannot be avoided, i.e. it is not possible in general to replace $T_{x,y}$ in (*) by another matrix which depends integrally on T_x and T_y because this would contradict the theorem of Hurwitz and Radon to be mentioned later.

SECOND PROOF (by Witt, using equivalence relations of binary quadratic forms). Here the inductive step for $\psi = \varphi \otimes \langle 1, a \rangle = \varphi \oplus a\varphi$ runs as follows: We have

$$\psi \cong \varphi \oplus a\varphi \cong \varphi(x)\varphi \oplus a\varphi(y)\varphi \cong \langle \varphi(x), a\varphi(y) \rangle \otimes \varphi$$

over the field $K(x,y)$. Now apply Lemma 1.3 with the element $\psi(x,y) = \varphi(x) + a\varphi(y)$ which is obviously nonzero in $K(x,y)$ and represented by $\langle \varphi(x), a\varphi(y) \rangle$. Then we may continue:

$$\psi \cong \psi(x,y)\langle 1, a\varphi(x)\varphi(y)\rangle \otimes \varphi \cong \psi(x,y)(\varphi \oplus a\varphi(x)\varphi(y)\varphi)$$
$$\cong \psi(x,y)(\varphi \oplus a\varphi) \cong \psi(x,y)\psi \quad \text{over } K(x,y).$$

This proves (2) for ψ instead of φ and finishes the proof.

Obviously the second proof is much shorter and more elegant. But the first proof also has its advantages. It is more elementary and explicit. Furthermore it implies the following useful result.

2.3 Corollary. The first row of T_x equals the row vector $x'A$. Therefore $\varphi(x)\varphi(y) = \varphi(z)$ has a solution z where the first component z_1 of z equals

$$z_1 = x'Ay = b_\varphi(x,y).$$

PROOF. The definition of $T_{x,y}$ immediately shows: If the first line of T_x equals $x'A$ then the first line of $T_{x,y}$ equals $(x'A, ay'A) = \left(\begin{smallmatrix} x \\ y \end{smallmatrix}\right)' B$. This proves the first claim by induction on k (for $\dim \varphi = 2^k$). The second claim is an immediate consequence of the first.

The most important case of 2.2 and 2.3 occurs for $a_1 = \ldots = a_k = 1$:

2.4 Corollary. For any 2-power $n = 2^k$ the n-fold unit form $n \times \langle 1 \rangle$ is strictly multiplicative. In other words there exist elements

$$z_i = \sum_{j=1}^{n} t_{ij}(x)y_j \quad (i = 1, \ldots, n)$$

with $t_{ij}(x) \in K(x) = K(x_1, \ldots, x_n)$ such that

$$(x_1^2 + \ldots + x_n^2)(y_1^2 + \ldots + y_n^2) = (z_1^2 + \ldots + z_n^2).$$

One can choose $t_{1j}(x) = x_j$, i.e. $z_1 = x_1 y_1 + \ldots x_n y_n$.

Historical Note. I proved 2.4 in [Pfister 1965$_1$]. On a suggestion of H. Lenz I generalized this to 2.2 and 2.3 in [Pfister 1965$_2$]. Witt found his proof in 1967. It was first published in [Lo]. Multiplicative forms are generalizations of the so-called *composition forms*, i.e. regular quadratic forms φ over K which satisfy equation (1) with a vector z whose components z_i are K-bilinear expressions in the x_j, y_k, say

$$z_i = \sum_{j,k=1}^{n} t_{ijk} x_j y_k \quad (i = 1, \ldots, n)$$

with constants $t_{ijk} \in K$. A famous theorem going back to Hurwitz (1898) states that such forms can exist only in dimensions 1, 2, 4, 8 provided char $K \neq 2$. They can actually be realized by the forms

$$\langle 1, a_1 \rangle \otimes \ldots \otimes \langle 1, a_k \rangle$$

for $k = 0, 1, 2, 3$. For details see [L] Chapter V, Theorem 5.10; [Shapiro 1984]; [R].

§3. Classification of Multiplicative Forms.
Consequences for $W(K)$

Using the results of section 2 it is not difficult to derive a full description of the multiplicative and strictly multiplicative forms.

3.1 Lemma. A (regular) quadratic form φ over K is multiplicative if and only if $D_L(\varphi)$ is a subgroup of $L^\bullet$ for every extension field L of K.

PROOF. See 1.10(3) of Chapter 1 for the definition of $D_L^{\cdot}(\varphi)$.

1) If φ is multiplicative over K then φ represents $\varphi(x)\varphi(y)$ over $K(x,y)$, hence over $L(x,y)$. Apply Proposition 3.1 of Chapter 1 to the polynomial $p = \varphi(x)\varphi(y) \in L[x,y]$. This shows that φ_L represents every element $\varphi(u)\varphi(v)$ where $u, v \in L^n$, $n = \dim\varphi$. For arbitrary $a = \varphi(u) \in D_L^{\cdot}(\varphi)$ and $b = \varphi(v) \in D_L^{\cdot}(\varphi)$ we get $ab \in D_L^{\cdot}(\varphi)$. Since $\frac{1}{a} = \varphi(\frac{u}{a})$ we also have $\frac{1}{a} \in D_L^{\cdot}(\varphi)$. Thus $D_L^{\cdot}(\varphi)$ is closed under multiplication and inversion. It is therefore a subgroup of $L^{\bullet}$.

2) Conversely, if $D_L^{\cdot}(\varphi)$ is a group for every field $L \supseteq K$ this holds in particular for the rational function field $L = K(x,y) = K(x_1,\ldots,x_n,y_1,\ldots,y_n)$ and shows that φ represents $\varphi(x)\varphi(y)$ over $K(x,y)$. Hence φ is multiplicative over K.

Corollary. Every multiplicative form φ over K represents all squares

$$a^2,\ 0 \neq a \in K,$$

or equivalently:

every multiplicative form represents 1.

The next theorem characterizes multiplicative and strictly multiplicative forms over a field K with char $K \neq 2$.

3.2 Theorem.

(1) Every anisotropic multiplicative form φ over K is of type $\varphi \cong \langle 1, a_1 \rangle \otimes \ldots \otimes \langle 1, a_k \rangle$, and hence strictly multiplicative.

(2) An isotropic form φ over K is always multiplicative. It is strictly multiplicative if and only if

$$\varphi \sim 0,\ \text{i.e. } \varphi \cong i \times \langle 1, -1 \rangle,\ i \geq 1.$$

PROOF.

(1) We have the following implications:
φ multiplicative $\Longrightarrow$ φ represents $\varphi(x)\varphi(y)$ over $K(x)(y)$ $\Longrightarrow$ φ contains $\varphi(x)\varphi$ over $K(x)$ by Theorem 3.4 of Chapter 1 (since φ remains anisotropic over $K(x)$) $\Longrightarrow$ $\varphi \cong \varphi(x)\varphi$ over $K(x)$ since both sides have the same dimension. Thus φ is strictly multiplicative. Let now $k \geq 0$ be the maximal integer such that φ contains a form $\psi \cong \langle 1, a_1 \rangle \otimes \ldots \otimes \langle 1, a_k \rangle$ over K. We claim: $\varphi \cong \psi$. Assume for a moment that $\varphi \cong \psi \oplus \chi$ with $\dim\chi \geq 1$, say $\chi \cong \langle b, \ldots \rangle$. Let z be an indeterminate vector of length 2^k. Then

$$\psi \oplus \chi \cong \varphi \cong \psi(z)\varphi \cong \psi(z)\psi \oplus \psi(z)\chi \cong \psi \oplus \psi(z)\chi$$

over $K(z)$ since φ and ψ are strictly multiplicative. From Witt's theorem we deduce $\chi \cong \psi(z)\chi$ over $K(z)$. In particular χ represents $b\psi(z)$ over $K(z)$. Applying the Subform Theorem 3.4 of Chapter 1 again we see that χ contains $b\psi$ whence φ contains $\psi \oplus b\psi \cong \psi \otimes \langle 1, b \rangle$ over K. This is a contradiction to the maximality of k. So we must have $\chi = 0$, $\varphi \cong \psi$.

(2) φ isotropic $\implies$ φ universal (over any field $L \supseteq K$) $\implies$ φ represents $\varphi(x)\varphi(y)$ over $K(x,y)$ $\implies$ φ is multiplicative over K.
$\varphi \cong i \times \langle 1, -1 \rangle \implies \varphi \cong \varphi(x)\varphi$ over $K(x)$, since $\langle 1, -1 \rangle \cong f(x)\langle 1, -1 \rangle$ for any nonzero element $f(x) \in K(x) \implies \varphi$ is strictly multiplicative.
Conversely consider an isotropic strictly multiplicative form φ over K. Let $\varphi \cong i \times \langle 1, -1 \rangle \oplus \varphi_0$ with $i \geq 1$, φ_0 anisotropic. We get

$$i \times \langle 1, -1 \rangle \oplus \varphi_0 \cong \varphi \cong \varphi(x)\varphi \cong i \times \varphi(x)\langle 1, -1 \rangle \oplus \varphi(x)\varphi_0$$

over $K(x)$. Cancelling $\langle 1, -1 \rangle \cong \varphi(x)\langle 1, -1 \rangle$ i times we conclude that $\varphi_0 \cong \varphi(x)\varphi_0$ over $K(x)$.
Assume $\varphi \not\sim 0$, i.e. $\varphi_0 \neq 0$, $\dim \varphi_0 \geq 1$, say $\varphi_0 = \langle b, \dots \rangle$. Then φ_0 represents $b\varphi(x)$. Since φ_0 is anisotropic we deduce that φ_0 contains $b\varphi$ which is impossible since $\dim \varphi > \dim \varphi_0$.

3.1 and 3.2 imply the following.

3.3 Corollary. If $D_K(n) = D_K(n \times \langle 1 \rangle)$ is closed under multiplication for *every* field K then n is a power of 2.

For particular fields the above property can of course hold without n being a power of 2. For example $D_{\mathbb{R}}^{\cdot}(n) = D_{\mathbb{R}}^{\cdot}(1)$ is a group for every $n \geq 1$.

Similarly, $D_K^{\cdot}(n) = D_K(4)$ for every $n \geq 4$ and every number field. (See Chapter 7.)

In the rest of this section we derive some consequences for the structure of the Witt ring $W(K)$ which depend on or are related to multiplicative forms. However, proofs are only given for those statements which are needed later, or are very elementary. The more interesting things are just stated with references to the literature.

First we consider the torsion elements in the Witt group $W(K)$. As for every abelian group the set $W_t(K) = \{w \in W(K) : \ell \times w = 0 \text{ for some } \ell \in \mathbb{N}\}$ is a subgroup of $W(K)$. For $w \in W_t(K)$ the smallest ℓ such that $\ell \times w = 0$ is called the order of w.

3.4 Theorem. $W_t(K)$ is a 2-group, i.e. the order of every element $w \in W_t(K)$ is a power of 2.

PROOF. Suppose $w' \in W_t(K)$ has order $\ell = 2^r k$ where k is odd and $k > 1$. Then $w = 2^r w'$ has odd order k. Represent $w = \tilde{\varphi}$ by the anisotropic quadratic form $\varphi = \langle a_1, \dots, a_m \rangle$. Then k is the smallest number such that $k \times \varphi \sim 0$.

Choose now any 2-power n with $n > m$ and consider the binary form $\psi = \langle 1, -\sum_1^n x_i^2 \rangle$ over $K(x) = K(x_1, \ldots, x_n)$ where the x_i are indeterminates over K. Then $n \times \psi$ is strictly multiplicative and isotropic over $K(x)$, hence $n \times \psi \sim 0$ over $K(x)$ by Theorem 3.2.

From $k \times \varphi \sim 0$ and $n \times \psi \sim 0$ we get $k \times (\varphi \otimes \psi) \sim 0$ and $n \times (\varphi \otimes \psi) \sim 0$. This implies $\varphi \otimes \psi \sim 0$ over $K(x)$ since 1 is a $\mathbb{Z}$-linear combination of the odd number k and the 2-power n. Finally we find $\varphi \cong (\sum_1^n x_i^2)\varphi$ over $K(x)$. In particular φ represents the element $a_1 \sum_1^n x_i^2$ over $K(x)$. $a_1 \sum_1^n x_i^2$ is the generic element represented by the quadratic form $n \times \langle a_1 \rangle$. As φ was anisotropic over K the subform theorem implies that φ contains $n \times \langle a_1 \rangle$ and $m = \dim \varphi \geq n$: contradiction.

As we have seen in section 1 the dimension and the determinant of a quadratic form φ are *invariants* in the sense that they remain unchanged if φ is replaced by an equivalent form $\psi \cong \varphi$. For the study of $W(K)$ it is necessary to find invariants which remain unchanged if φ is replaced by a similar form $\psi \sim \varphi$. It suffices to test the form $\psi = \varphi \oplus \langle 1, -1 \rangle$. Obviously the dimension and the determinant are no longer invariants but the parity of the dimension and the discriminant are invariants of the similarity class $\tilde{\varphi}$. The equations $\dim(\varphi \oplus \psi) = \dim \varphi + \dim \psi$, $\dim(\varphi \otimes \psi) = \dim \varphi \cdot \dim \psi$ imply the corresponding congruences modulo 2. This leads to the following.

3.5 Definition.

(1) $e_0(\tilde{\varphi}) = \dim \varphi \bmod 2$ is called the *dimension index* of a similarity class $\tilde{\varphi} \in W(K)$. The map $e_0 : W(K) \to \mathbb{Z}/2\mathbb{Z}$ is a surjective ring-homomorphism.

(2) $I(K) = \ker e_0$ is called the *fundamental ideal* of $W(K)$. $I(K)$ consists of the (classes of) even-dimensional forms.

The factor group $W(K)/I(K)$ is isomorphic to $\mathbb{Z}/2\mathbb{Z}$ via e_0.

The discriminant $d(\tilde{\varphi}) = d(\varphi)$ defines a well-defined map from $W(K)$ to the square class group $G(K)$. Unfortunately, we do not have $d(\varphi \oplus \psi) = d(\varphi) \cdot d(\psi)$ in general. But if we restrict to forms φ of even dimension then $d(\varphi) = (-1)^{\frac{\dim \varphi}{2}} \det \varphi$ by the definition of the discriminant. Therefore $d(\tilde{\varphi} \oplus \tilde{\psi}) = d(\tilde{\varphi}) \cdot d(\tilde{\psi})$ if $\tilde{\varphi}, \tilde{\psi} \in I(K)$. We get

3.6 Proposition

(1) The map $e_1 : I(K) \to G(K)$ with $e_1(\tilde{\varphi}) = d(\tilde{\varphi})$ for $\tilde{\varphi} \in I(K)$ is a surjective group-homomorphism.

(2) $\ker e_1 = I^2$, the square of the fundamental ideal.

(3) The factor group I/I^2 is isomorphic to $G(K)$ via e_1.

PROOF.

(1) e_1 is surjective since $d(\langle 1, -a \rangle) = aK^{\bullet 2} \in G(K)$ for any $a \in K^{\bullet}$.

(2) As a group $I = I(K)$ is generated by the binary forms $\langle a, b \rangle$. Thus I^2 is generated by the (classes of) 4-dimensional forms

$$\varphi = \langle a, b \rangle \otimes \langle c, d \rangle = \langle ac, ad, bc, bd \rangle.$$

We see immediately that $d(\varphi) = \det \varphi = 1 \in G(K)$. Hence I^2 lies in the kernel of e_1.

Conversely let $\tilde{\varphi} \in I$ be such that $e_1(\tilde{\varphi}) = 1$.

$\tilde{\varphi}$ is represented by a quadratic form $\varphi = \langle a_1, \ldots, a_{2n} \rangle$ of even dimension $2n$. We use induction on n. For $n = 1$ we have $\varphi = \langle a_1, a_2 \rangle$ with $a_1 a_2 \in K^{\bullet 2}$. Hence $a_2 = -a_1$ (up to a square factor), $\varphi \cong a_1 \langle 1, -1 \rangle$, $\tilde{\varphi} = 0$ in $W(K)$.

The inductive step $n - 1 \to n$ goes as follows: Write $\varphi = \langle a_1, a_2, a_3 \rangle \oplus \langle a_4, \ldots, a_{2n} \rangle$, $\varphi \sim \langle a_1, a_2, a_3, a_1 a_2 a_3 \rangle \oplus \langle -a_1 a_2 a_3, a_4, \ldots, a_{2n} \rangle$.
Here $\varphi_1 = \langle a_1, a_2, a_3, a_1 a_2 a_3 \rangle \cong \langle a_1, a_2 \rangle \otimes \langle 1, a_1 a_3 \rangle \in I^2$,
and $\varphi_2 = \langle -a_1 a_2 a_3, a_4, \ldots, a_{2n} \rangle$ has dimension $2(n-1)$ and discriminant $d(\varphi_2) = \frac{d(\varphi)}{d(\varphi_1)} = \frac{1}{1} = 1$. Hence $\tilde{\varphi}_2 \in I^2$ by the induction hypothesis. This proves (2).

(3) is an immediate consequence of (1) and (2).

It is interesting to note that the subgroup $I(K)$ of $W(K)$ is generated not only by the general binary forms $\langle a, b \rangle$ but already by the multiplicative binary forms $\langle 1, a \rangle$, $a \in K^{\bullet}$. This follows from the relation $\langle a, b \rangle \sim \langle 1, a \rangle \ominus \langle 1, -b \rangle$, and the equation $\widetilde{\langle a, b \rangle} = \widetilde{\langle 1, a \rangle} \ominus \widetilde{\langle 1, -b \rangle}$, in $W(K)$. More generally we have the following.

3.7 Observation. For every $n \in \mathbb{N}$ the n-th power I^n of the fundamental ideal $I \subset W$ is generated (as an additive group) by the strictly multiplicative forms $\varphi = \langle 1, a_1 \rangle \otimes \ldots \otimes \langle 1, a_n \rangle$, $a_i \in K^{\bullet}$ $(i = 1, \ldots, n)$.

It is this observation which somehow "explains" the importance of the multiplicative forms for the structure theory of $W(K)$.

It is convenient to introduce the following

Notation. $\ll a_1, \ldots, a_n \gg := \langle 1, a_1 \rangle \otimes \ldots \otimes \langle 1, a_n \rangle$.

Regarding 3.5 and 3.6 it is quite natural to study the quotient groups $\bar{I}^n = I^n / I^{n+1}$ for all $n \geq 0$ or even the "graded Witt ring"

$$\hat{W} = \bar{I}^0 \oplus \bar{I}^1 \oplus \bar{I}^2 \oplus \ldots$$

where the multiplication $\bar{I}^m \times \bar{I}^n \to \bar{I}^{m+n}$ is induced by the tensor product

$$I^m \times I^n \ni (\varphi, \psi) \mapsto \varphi \otimes \psi \in I^{m+n}.$$

Without proof I state some main results and conjectures which have been obtained during the last 30 years:

A) For every regular quadratic space (V, φ) over K of even dimension n its Clifford algebra $C(\varphi)$ is a central simple K-algebra of dimension 2^n. The class $c(\varphi)$ of $C(\varphi)$ in the Brauer group $B(K)$ of K is an invariant of the similarity class $\tilde{\varphi} \in W(K)$ and it has order ≤ 2 in $B(K)$. (For historical reasons this invariant is often called the (Minkowski–)Hasse–Witt invariant.) See [O'M], [L], [S].

B) $e_2 = c|_{I^2}$ is a group homomorphism $e_2 : I^2 \to B_2(K)$ with $I^3 \subseteq \ker e_2$. Hence e_2 induces a homomorphism $\bar{e}_2 : \bar{I}^2 \to B_2(K)$.
   Here $B_2(K) = \{c \in B(K) : 2c = 0\}$.
   Merkurjev [1981] has shown that $\bar{e}_2$ is an isomorphism. For simpler proofs see [Arason 1984], [Kersten 1990].

C) The seemingly unrelated target groups $\mathbb{Z}/2\mathbb{Z}$, $G(K)$, $B_2(K)$ of the homomorphisms e_0, e_1, e_2 are in fact (naturally isomorphic to) the cohomology groups $H^n = H^n(K) := H^n(\Gamma, \mathbb{Z}/2\mathbb{Z})$ for the cases $n = 0, 1, 2$ where Γ is the absolute Galois group of the field K (i.e. the automorphism group of K_s/K for a separable closure K_s of K), operating trivially on $\mathbb{Z}/2\mathbb{Z}$.
   It is therefore natural to expect *higher invariants*, that is group homomorphisms $e_n : I^n \to H^n$ for all n. Milnor [1970] asked whether e_n exists and $\bar{e}_n$ is an isomorphism for all n and all fields (**Milnor conjecture**).

D) Up to now the following have been proved:
   e_3 exists [Araason 1975].
   e_4 exists, $\bar{e}_3$ is an isomorphism [Jacob–Rost 1989 and independently Merkurjev–Suslin 1990/1991].
   e_5 exists, $\bar{e}_4$ is an isomorphism [Rost, unpublished]. Furthermore, Arason, Elman and Jacob proved in a series of papers [see the survey article 1989] that for fields of low transcendence degree over the reals the full Milnor conjecture holds.

E) For every field K with char $K \neq 2$ we have the following.

   Intersection Theorem [Arason–Pfister 1971]

   $$\bigcap_{n \in \mathbb{N}} I^n(K) = 0.$$

Together with the Milnor conjecture this would imply that the invariants $e_0, e_1, e_2, \ldots$ form a complete system of invariants for quadratic forms, at least in a weak sense.

§4. Appendix: The case char $K = 2$

As we have seen in Chapter 1 it is necessary to distinguish between symmetric bilinear and quadratic forms if char $K = 2$. We shall first consider the symmetric bilinear forms

$$b(x,y) = x'Ay \quad \text{with } A = A' \in M_n(K).$$

As long as there exists a vector $v \in V = K^n$ with $b(v,v) \neq 0$ we get an orthogonal decomposition

$$V = Kv \oplus (Kv)^\perp.$$

If $b(v,v) = 0$ for every v but if there exists a pair of (necessarily independent) vectors $u,v \in V$ such that $b(u,v) \neq 0$ we get a splitting

$$V = (Ku + Kv) \oplus (Ku + Kv)^\perp.$$

On normalizing $b(u,v) = 1$ we find that $(Ku + Kv, b)$ is equivalent to the *hyperbolic plane* H with matrix $\left(\begin{smallmatrix}0&1\\1&0\end{smallmatrix}\right)$ as a bilinear space. This proves

4.1 Theorem. Every symmetric bilinear space (V,b) over K has an orthogonal decomposition

$$(V,b) \cong (a_1) \oplus \ldots \oplus (a_r) \oplus i \times H \oplus \text{rad } (V,b)$$

where (a_j) is the 1-dimensional bilinear space Ke_j with $b(e_j,e_j) = a_j \neq 0$, where $i \geq 0$ and where rad $(V,b) := \{w \in V : b(w,v) = 0 \text{ for all } v \in V\}$ is the radical of the symmetric bilinear space (V,b).

From now on all bilinear spaces will be regular, i.e. rad $(V,b) = 0$ or equivalently det $A \neq 0$. This is no serious restriction.

Contrary to the quadratic case there are several regular isotropic symmetric bilinear forms b of dimension 2 (b is called isotropic if $b(x,x) = 0$ has a nontrivial solution over K), namely H and the forms $M_a \cong (a) \oplus (-a) \cong (a) \oplus (a)$, $a \in G(K) = K^\bullet/K^{\bullet 2}$. Since $ax_1^2 - ax_2^2 = a(x_1 + x_2)^2$ the square class of a is the only value represented by M_a which shows that the spaces M_a are pairwise non-isometric.

Notation. M_a is called *metabolic* or *split plane*. Clearly H is not isometric to any M_a.

The cancellation law does not hold for regular symmetric bilinear forms as is easily shown by

4.2 Example.

$$(a) \oplus (a) \oplus (a) \cong (a) \oplus H$$

and
$$M_a \oplus M_a \cong M_a \oplus H \quad \text{for } a \in G(K).$$

In order to define the Witt group of regular symmetric bilinear forms over K it is therefore necessary to define $M_a \sim 0$ for any $a \in G(K)$ and $H \sim 0$ (since the cancellation law holds in every group).

4.3 Definition.

(1) A regular symmetric bilinear space (V, b) over K is called *metabolic* (or split) if it is an orthogonal sum of isotropic planes, i.e. spaces M_a or H. We then write
$$(V, b) \sim 0.$$

(2) Two (regular) symmetric bilinear spaces $(U_1, b_1), (U_2, b_2)$ over K are called *similar*, $(U_1, b_1) \sim (U_2, b_2)$, if there exist metabolic spaces (V, b), (V', b') such that

$$(U_1, b_1) \oplus (V, b) \cong (U_2, b_2) \oplus (V', b').$$

(3) The set of similarity classes of regular symmetric bilinear spaces over K together with the operations $\oplus$ (orthogonal sum) and $\otimes$ (tensor product) is called the *Witt ring* $W(K)$ of K.

Notes.

(1) From 4.1 we see that every element $\neq 0$ of $W(K)$ may be represented by an *anisotropic* diagonal space

$$(V, b) \cong (a_1) \oplus \ldots \oplus (a_r).$$

It is not hard to show that (V, b) is unique up to isometry.

(2) Since $(a) \oplus (a) \cong M_a \sim 0$ every element of $W(K)$ is its own inverse with respect to $\oplus$, or in other words

$$2w = 0 \quad \text{for every } w \in W(K).$$

(3) The multiplication on $W(K)$ is induced by $(a_1) \otimes (a_2) = (a_1 a_2)$. A coordinate-free definition is as follows: $(V_1, b_1) \otimes (V_2, b_2) = (V_1 \otimes V_2, b_1 \otimes b_2)$ with

$$(b_1 \otimes b_2)(\sum_i u_{1i} \otimes u_{2i}, \sum_j v_{1j} \otimes v_{2j}) = \sum_{i,j} b_1(u_{1i}, v_{1j}) b_2(u_{2i}, v_{2j})$$

for $u_{1i}, v_{1j} \in V_1$, $u_{2i}, v_{2j} \in V_2$ and finite sums over i, j.

The dimension index $e_0(b) = \dim b \bmod 2$ and the discriminant (= determinant) $e_1(b) = d(b) \in G(K)$ are invariants of the Witt class $w = \tilde{b}$ of a regular symmetric bilinear form b (as in the case char $K \neq 2$). For more details about $W(K)$, especially in the case where K has finite dimension over its subfield K^2, see [Milnor 1971].

We shall now turn to quadratic forms. As seen in Proposition 4.5 of Chapter 1 there is only one regular isotropic quadratic space of dimension 2, the *hyperbolic plane* $H_q \cong [0, 1, 0]$. (The suffix q stands for quadratic, H_q should not be confused with the hyperbolic plane H as a symmetric bilinear space.) Despite Example 4.4 of Chapter 1 we have the following important result.

4.4 Theorem. Witt's cancellation theorem holds for *regular* quadratic spaces over a field of characteristic 2, i.e. $\varphi \oplus \varphi_1 \cong \varphi \oplus \varphi_2$ with regular quadratic forms $\varphi, \varphi_1, \varphi_2$ over K implies $\varphi_1 \cong \varphi_2$.

PROOF. See [Arf 1941], [S, Ch. IX §4], [MH, App. 1] or [B, Ch. III §4].

It is an immediate consequence of this theorem that similarity $\sim$ and the set $Wq(K)$ of similarity classes of regular quadratic forms over K may be introduced as in the case char $K \neq 2$. Clearly every element in $Wq(K)$ is represented by a unique (up to isometry) anisotropic quadratic form φ_0. $Wq(K)$ is an abelian group under the operation $\oplus$ (= direct orthogonal sum).

4.5 Example. Every binary $\varphi = [a, 1, c]$ satisfies $\varphi \oplus \varphi \sim 0$, hence $2w = 0$ for every $w \in Wq(K)$.

PROOF. Let e_1, e_2, e_3, e_4 be the standard basis for $(V, \varphi \oplus \varphi)$. Consider the two subspaces U_1 generated by $e_1, e_2 + e_4$ and U_2 generated by $e_1 + e_3, e_4$. Then $U_1 \cong [a, 1, 0] \cong H_q$, $U_2 \cong [0, 1, c] \cong H_q$ and $V \cong U_1 \oplus U_2$.

Since all regular quadratic forms have even dimension the dimension index $e_0(\tilde{\varphi}) = \dim \varphi \bmod 2$ is trivial on $Wq(K)$. The discriminant has to be replaced by the so-called *Arf invariant* $\Delta(\varphi)$.

4.6 Definition.

(1) For a field K of characteristic 2 let $\mathcal{P}(K) = \{x^2 + x : x \in K\}$ (this is a subgroup of $(K, +)$) and let
$S(K) = K/\mathcal{P}(K)$ be the additive group of separable field extensions of degree 2 over K (for $b \in S(K)$ the field $L = K(x)$ is given by the equation $x^2 + x + b = 0$).

(2) For a regular quadratic form

$$\varphi \cong \bigoplus_{i=1}^{r} [a_i, 1, c_i]$$

let $\Delta(\tilde{\varphi}) = \Delta(\varphi) = \sum_{i=1}^{r} a_i c_i \in S(K)$. This element is called the Arf invariant of φ or $\tilde{\varphi}$.

Of course it has to be shown that $\Delta(\varphi)$ is well-defined and invariant under similarity. See [Arf 1941] or [S, Ch. IX §4].

4.7 Examples.

(1) K quadratically closed, char $K = 2$. Then $G(K) = \{1\}, S(K) = 0$ and $|W(K)| = 2, Wq(K) = 0$.

(2) $K = \mathbf{F}_q$ finite field of characteristic 2 (q = power of 2). Then $K = K^2, |S(K)| = 2$ and $|W(K)| = |Wq(K)| = 2$.

The connection between (regular) symmetric bilinear forms and quadratic forms over a field of characteristic 2 is given by

4.8 Proposition. $Wq(K)$ is a $W(K)$-module via the rules

$$W(K) \times Wq(K) \ni (b, \varphi) \mapsto \Phi \in Wq(K)$$

with

$$
\begin{aligned}
\Phi(x \otimes y) &= b(x, x)\varphi(y), \\
b_\Phi(x_1 \otimes y_2, x_2 \otimes y_2) &= b(x_1, x_2)b_\varphi(y_1, y_2).
\end{aligned}
$$

(The underlying vector-space of Φ is the tensor product of the underlying spaces of b and φ. Usually Φ is therefore denoted by $\Phi = b \otimes \varphi$.)

PROOF. See [S, Ch.I §6], [MH, App.1], [B, Ch.I] or [Sah 1962].

For regular quadratic forms over a field K with char $K = 2$ the notions of multiplicative and strictly multiplicative quadratic forms may be defined as in 2.1.

Since there are no regular quadratic forms of dimension 1 the lowest dimension for a regular multiplicative form φ is 2. Also, φ must represent 1, hence

$$\varphi \cong \varphi_a \cong [1, 1, a].$$

This form is indeed (strictly) multiplicative and even a composition form since it is the norm form of the separable quadratic extension $L = K(\alpha)$ with

$$\alpha^2 + \alpha + a = 0.$$

An element $x = x_1 + x_2\alpha$ ($x_1, x_2 \in K$) has conjugate

$$\bar{x} = x_1 + x_2(\alpha + 1) \quad \text{and norm} \quad Nx = x\bar{x} = x_1^2 + x_1 x_2 + ax_2^2 = \varphi(x).$$

Hence $\varphi(x)\varphi(y) = \varphi(xy)$ for $x, y \in L$.

Let us now study the 2^n-dimensional quadratic forms

$$\varphi = \ll a_1, \ldots, a_{n-1}, a_n \gg := (1, a_1) \otimes \ldots \otimes (1, a_{n-1}) \otimes [1, 1, a_n]$$

where $n \geq 1$, $a_n \in K$, $a_1, \ldots, a_{n-1} \in K^{\bullet}$ and where the forms $(1, a_i)$ are considered as regular symmetric bilinear forms. With φ instead of $\ll a_1, \ldots, a_n \gg$ Theorems 2.2 and 3.2 hold for char $K = 2$ and the proofs can be taken over nearly verbatim. (Note that the Subform Theorem also holds, see 4.9 of Chapter 1.)

A strictly multiplicative quadratic form φ as above is a composition form if and only if $n \leq 3$. This was proved by [Albert 1942] in complete analogy to the case char $K \neq 2$. In the same paper he found the following interesting new phenomenon about multiplicative quadratic forms in characteristic 2 if we do not require regularity.

Let $\varphi = \langle 1, a_1 \rangle \otimes \ldots \otimes \langle 1, a_n \rangle$ with $a_i \in K^{\bullet}$ be an *anisotropic* non-defective diagonal quadratic form. φ corresponds to the multi-quadratic field extension $L = K(\sqrt{a_1}, \ldots, \sqrt{a_n})$ which is purely inseparable of degree 2^n. The generic element $x \in L$ has the form

$$x = \sum_{I \subseteq \{1, \ldots, n\}} x_I \alpha_I$$

where $x_I \in K$, $\alpha_\emptyset = 1$, $\alpha_I = \sqrt{a_{i_1} \ldots a_{i_r}}$ for $I = \{i_1, \ldots, i_r\}$, $1 \leq i_1 < i_2 < \ldots < i_r \leq n$. We get $\varphi(x) = x^2 = \sum_{I \subseteq \{1, \ldots, n\}} a_I x_I^2$, $a_I = \alpha_I^2$. This shows $\varphi(x)\varphi(y) = \varphi(xy)$, i.e. φ is a composition form for every $n \in \mathsf{N}$! (Of course it depends on the field K for which $n \in \mathsf{N}$ a form φ as above can possibly be anisotropic.)

Finally we comment on the structure of $W(K)$ and $Wq(K)$ and the analogue of the Milnor conjecture for a field K of characteristic 2. For $W(K)$ we get

$$\begin{aligned} I &= \{\tilde{b} \in W(K) : e_0(b) = 0\}, \\ I^2 &= \{\tilde{b} \in I : e_1(b) = 1\} \quad \text{and} \quad \bigcap_n I^n = 0. \end{aligned}$$

(For a proof see [Arason–Pfister 1971].)

For Wq one finds $I_q = \ker \Delta = I \otimes Wq \subset W \otimes Wq = Wq$. Further results on W and Wq may be found in [Arason 1979].

The analogue of the Milnor conjecture reads as follows:
Let Ω^n be the n-th exterior power of the absolute differential module $\Omega = \Omega^1_{K/\mathbf{z}}$. Consider the homomorphism

$$\nu : \Omega^n \to \Omega^n/d\Omega^{n-1}, \quad x\frac{dy_1}{y_1} \wedge \ldots \wedge \frac{dy_n}{y_n} \mapsto (x^2 + x)\frac{dy_1}{y_1} \wedge \ldots \wedge \frac{dy_n}{y_n}.$$

Let $\nu(n)$ be the kernel of ν, H^{n+1} the cokernel of ν (the classical cohomology groups as defined at the end of section 3 cannot be used here since they vanish for $n > 1$ if char $K = 2$). Then we have

F)   **Theorem of Kato.**

$$I^n/I^{n+1} \cong \nu(n) \quad \text{for all } n \geq 0,$$
$$I^n Wq/I^{n+1}Wq \cong H^{n+1} \quad \text{for all } n \geq 0.$$

PROOF. See [Kato 1982].

This leads to the remarkable fact that we know more about the case char $K = 2$ than about the case char $K \neq 2$!

Chapter 3

The Level of Fields, Rings, and Topological Spaces

§1. The Level of Fields

In a fundamental paper of E. Artin and O. Schreier it was shown that the algebraic investigation of (commutative) fields requires us to distinguish between formally real and nonreal fields. The (formally) real fields are those which admit at least one linear ordering such as $\mathbf{Q}$ or $\mathbf{R}$. They will play a role in Chapter 6. In the present chapter we concentrate on nonreal fields. We start with

1.1 Definition.

(1) A field K is called *nonreal* if -1 is a sum of (finitely many) squares in K. The number

$$s = s(K) := \min\{n : -1 = e_1^2 + \ldots + e_n^2 \quad \text{with } e_i \in K\}$$

is called the *level of* K.

(2) A field K is called (formally) *real* if -1 is not a sum of squares in K. In this case we put $s(K) = \infty$.

Note. The letter s stems from the German word "Stufe" for the level. The French word is "niveau".

1.2 Examples of real and nonreal fields.

(1) Every subfield of $\mathbf{R}$ is real.

(2) $s(\mathbf{C}) = 1$ since $-1 = i^2$ in $\mathbf{C}$.

(3) char $K = 2 \Longrightarrow -1 = 1$ in $K \Longrightarrow s(K) = 1$.

(4) $K = \mathbf{F}_q$ finite field with q elements (q odd). Then

$$s(\mathbf{F}_q) = \begin{cases} 1 & \text{for} \quad q \equiv 1 \bmod 4, \\ 2 & \text{for} \quad q \equiv -1 \bmod 4 \end{cases}.$$

(5) K local field with residue field k (not necessarily finite), char $k \neq 2 \Longrightarrow$

$$s(K) = s(k).$$

(6) K dyadic local field (in the classical sense, i.e. the residue field k is finite of characteristic 2) $\implies$

$$s(K) = \begin{cases} 1 & \text{for} \quad \sqrt{-1} \in K, \\ 2 & \text{for} \quad [K : \mathbf{Q}_2] \text{ even}, \sqrt{-1} \notin K, \\ 4 & \text{for} \quad [K : \mathbf{Q}_2] \text{ odd.} \end{cases}$$

(7) K number field $\implies$

$$s(K) = \max\{s(K_p) : p \text{ place of } K\} \in \{1, 2, 4, \infty\}.$$

(8) $K \subseteq L \implies s(L) \le s(K)$.

(9) For any field K we have

$$s(K) = s(K(t)) = s(K((t))).$$

PROOF.

(1) –(3) and (8) are trivial.

(4) is easy, compare Example 1.10(3) of Chapter 2.

(5) follows from Hensel's Lemma.

(6) In the field $\mathbf{Q}_2$ of 2-adic numbers every 2-adic integer which is congruent to 1 mod 8 is a square. Then

$$-1 = 6 - 7 = 1^2 + 1^2 + 2^2 + (\sqrt{-7})^2.$$

This shows $s(\mathbf{Q}_2) \le 4$. On the other hand any integral square in $\mathbf{Q}_2$ is $\equiv$ 0, 1 or 4 mod 8. Thus any sum of three integral squares is $\equiv$ 0, 1, 2, 3, 4, 5 or 6 mod 8. Hence -1 is not a sum of three squares in $\mathbf{Q}_2$. (Show first that $-1 = e_1^2 + e_2^2 + e_3^2$ would imply $e_1, e_2, e_3 \in \mathbf{Z}_2$ since denominators cannot cancel.) A well-known theorem of T. Springer gives the result $s(K) = 4$ whenever $[K : \mathbf{Q}_2]$ is odd. The rest is easy.

(7) Apply the theorem of Minkowski and Hasse to the "local" results in (1), (2), (5) and (6).

(9) For $K(t)$ this follows from Lemma 2.1 of Chapter 1. The proof for the field $K((t))$ of formal Laurent (power) series over K is similar.

For more details, especially those concerning parts (5)–(7), see [L, Chapter 11].

Historical Note. The result (7) was claimed by Hilbert (1902) and first proved by [Siegel 1919] just a few years before the Hasse–Minkowski theorem.

The preceding examples motivate a question which was asked by v.d. Waerden in the early 1930s: What are the possible values for the level s of a nonreal field? H. Kneser [1934] investigated this question a little further and proved $s = 1, 2, 4, 8$ or a multiple of 16. But the main questions remained open: Is there a nonreal field K with $s(K) > 4$? Is $s(K)$ always a power of 2?

Applying the main results of Chapters 1 and 2 we can now easily answer these two questions.

1.3 Theorem. The level $s = s(K)$ of a nonreal field K is a power of 2.

PROOF. Suppose $2^m \leq s < 2^{m+1}$. We have to show $s = 2^m$. Put $2^m = n$, $\varphi = n \times \langle 1 \rangle = \underbrace{\langle 1, 1 \rangle \otimes \ldots \otimes \langle 1, 1 \rangle}_{m}$. By assumption we have

$$0 = 1 + e_1^2 + \ldots + e_{n-1}^2 + e_n^2 + \ldots + e_s^2$$

for suitable elements $e_i \in K$. Put

$$a = 1 + e_1^2 + \ldots + e_{n-1}^2, \quad b = e_n^2 + \ldots + e_s^2.$$

Then $a, b \in D_K(\varphi)$ and $a \neq 0$ since otherwise $s(K)$ would be less than n. φ being multiplicative we get

$$ab = c_1^2 + \ldots + c_n^2 \quad \text{for suitable } c_i \in K.$$

Now the equation $0 = a + b$ implies $-a^2 = ab$ and

$$-1 = \frac{ab}{a^2} = \left(\frac{c_1}{a}\right)^2 + \ldots + \left(\frac{c_n}{a}\right)^2.$$

Hence we must have $s = n$.

1.4 Theorem. Let k be a real field, let $2^m \leq n < 2^{m+1}$ for some $m \in \mathsf{N}_0$, let $K = k(x_1, \ldots, x_n)$ and

$$d = x_1^2 + \ldots + x_n^2 \in K.$$

Then the field $L = K(\sqrt{-d})$ has level $s(L) = 2^m$.

PROOF.

(1) Since $0 = (\sqrt{-d})^2 + x_1^2 + \ldots + x_n^2$ in L we have $s(L) \leq n$. By Theorem 1.3 we get $s(L) \leq 2^m$.

(2) Put $t := 2^m$ and assume $s = s(L) < t$. We have to derive a contradiction. The assumption $s < t$ implies the existence of elements $\alpha_i \in L$ ($i = 1, \ldots, t$), not all zero (say $\alpha_{s+1} = 1, \alpha_{s+2} = \ldots = \alpha_t = 0$), such that

(*)
$$\alpha_1^2 + \ldots + \alpha_t^2 = 0.$$

We have $\alpha_i = a_i + b_i \sqrt{-d}$ with $a_i, b_i \in K$. Here we must have $\sum_1^t b_i^2 \neq 0$. Otherwise all b_i would have to vanish since K is real (see Example 1.2(9), applied n times), we would get $\sum_1^t a_i^2 = 0$ and $a_i = 0$ for all i, hence $\alpha_i = 0$ for all $i = 1, \ldots, t$.

Since K is real and L is nonreal we have $\sqrt{-d} \notin K$. From (*) we therefore get two equations in K:

$$\sum_1^t a_i^2 - d \sum_1^t b_i^2 = 0, \quad \sum_1^t a_i b_i = 0.$$

We can now apply Corollary 2.4 of Chapter 2 and find

$$\sum_1^t a_i^2 \cdot \sum_1^t b_i^2 = \left(\sum_1^t a_i b_i \right)^2 + c_2^2 + \ldots + c_t^2 = \sum_{j=2}^t c_j^2$$

for suitable $c_2, \ldots, c_t \in K$. $d = \sum a_i^2 / \sum b_i^2$ is obtained on dividing the above equation by the square of the element $c = \sum b_i^2 \neq 0$. This leads to

$$d = x_1^2 + \ldots + x_n^2 = \sum_{j=2}^t \left(\frac{c_j}{c} \right)^2,$$

a sum of $t - 1$ squares in K. But $t - 1 = 2^m - 1 \leq n - 1 < n$. This is a contradiction to Corollary 3.3 of Chapter 1.

Note. The construction of fields with prescribed 2-power level does not say anything about the other question: What is the level of a prescribed field, say an explicitly given algebraic extension M of $K = k(x_1, \ldots, x_n)$? This is a much more difficult question. In general it will depend heavily on the individual field M. The only general statement that I know of is the following: If k is real closed (say $k = \mathbf{R}$) and M is nonreal then $s(M) \leq 2^n$. We shall prove this result in Chapter 6.

§2. The Level of Rings

The definition of the level applies not only for fields but for any ring R with identity element $1 \neq 0$:

2.1 Definition. The number

$$s(R) := \min\{n : -1 = e_1^2 + \ldots + e_n^2 \, , \ e_i \in R\}$$

is called the *level of R*. If -1 is not a sum of squares in R we define $s(R) = \infty$.

For the commutative rings turning up in number theory the results do not differ very much from the field case. Let us mention the following.

2.2 Examples.

(1) The ring $R = \mathbf{Z}/4\mathbf{Z}$ has level $s(R) = 3$.

(2) The ring R of integers of a p-adic number field has level $s(R) = 1, 2$ or 4.

(3) For an order R of a nonreal algebraic number field K we have $s(R) \leq 4$ but not always $s(R) = s(K)$. For the maximal order R of K we have $s(R) = s(K)$ in the cases $s(K) = 1$ or 4. If $s(K) = 2$ then $s(R) = 2$ or 3.

PROOF.

(1) is clear since $-1 = 1 + 1 + 1$ in R and $R^2 = \{0, 1\}$.

(2) is proved in [Riehm 1964].

(3) is proved in [Peters 1972].

Since we do not specialize in number theoretic aspects there is no point in repeating the proofs here. Similarly we do not consider noncommutative rings here though quite a number of research papers concerning the level of skew-fields and other noncommutative rings have appeared during the last decade. I refer the interested reader to the survey article "On the level" by [D.W. Lewis 1987] and the papers [Leep–Tignol–Vast 1989], [Leep 1990$_2$], [Denert–Tignol–Van Geel–Vast 1990] in the list of references.

The great surprise however was the discovery of commutative rings, actually even integral domains, with prescribed finite level by Dai, Lam and Peng [1980]. This is deeply connected to a "level-theory" for topological spaces with involution which was started by Dai and Lam [1984]. The main result is

2.3 Theorem. For every natural number n the integral domain

$$A_n := \mathbf{R}[x_1, \ldots, x_n]/(1 + x_1^2 + \ldots + x_n^2)$$

has level $s(A_n) = n$.

PROOF. The main ingredients of the proof are Theorem 3.7 in the next section and the Borsuk–Ulam theorem for which I shall offer a proof in the next chapter which presupposes only a minimum of topology. It remains to outline the elementary starting points of the proof.

(1) It is well-known that $\mathbf{R}[x_1,\ldots,x_n]$ is a unique factorization domain (see e.g. [Lang 1985 Chapter V]) and it is readily shown that $1+x_1^2+\ldots+x_n^2$ is a prime element since the only possible factorization would be of type $(x_n+c)(x_n-c)$ with $c\in\mathbf{R}[x_1,\ldots,x_{n-1}]$. Hence $(1+x_1^2+\ldots+x_n^2)$ is a prime ideal and A_n is an integral domain. Let $\bar{x}_i$ be the image of x_i under the natural epimorphism $\mathbf{R}[x_1,\ldots,x_n]\to A_n$. Then

$$-1=\bar{x}_1^2+\ldots+\bar{x}_n^2$$

which shows $s(A_n)\le n$.

(2) Let now $A(n)$ be any $\mathbf{R}$-algebra of exact level n. (The existence of such an algebra is shown in Theorem 3.7.) Let $-1=a_1^2+\ldots+a_n^2$ in $A(n)$. Consider the $\mathbf{R}$-algebra-homomorphism h from $\mathbf{R}[x_1,\ldots,x_n]$ to $A(n)$ sending 1 to 1 and x_i to a_i. h factorizes through the principal ideal $(1+x_1^2+\ldots+x_n^2)$ and induces an $\mathbf{R}$-algebra-homomorphism $\bar{h}:A_n\to A(n)$. Suppose now that $s(A_n)<n$, say $-1=e_1^2+\ldots+e_m^2$ in A_n with $m<n$. Application of $\bar{h}$ yields

$$-1=\bar{h}(e_1)^2+\ldots+\bar{h}(e_m)^2\quad\text{in }A(n),$$

contradicting the assumption $s(A(n))=n$. Hence $s(A_n)=n$.

(3) The explanation for part (2) of the proof is the fact that A_n is the *generic* $\mathbf{R}$-algebra which possibly has level n, i.e. it allows a homomorphism into any $\mathbf{R}$-algebra of level n if such an algebra exists. It is interesting to note that A_n has quotient field

$$K_n=\mathbf{R}(x_1,\ldots,x_{n-1},\sqrt{-(1+x_1^2+\ldots+x_{n-1}^2)}).$$

Adjoin a further indeterminate y_n to K_n and put $y_i=x_iy_n$ for $i=1,\ldots,n-1$. Then

$$
\begin{aligned}
K_n(y_n) &= \mathbf{R}(x_1,\ldots,x_{n-1},y_n,y_n\sqrt{-(1+x_1^2+\ldots+x_{n-1}^2)})\\
&= \mathbf{R}(y_1,\ldots,y_n,\sqrt{-(y_1^2+\ldots+y_n^2)})
\end{aligned}
$$

is exactly the field L of Theorem 1.4 (for $k=\mathbf{R}$). Thus

$$s(K_n)=s(K_n(y_n))=s(L)=2^m$$

where 2^m is the largest power of 2 which is smaller than or equal to n. In particular $s(K_n)<s(A_n)$ unless n is a power of 2.

§3. The Level of Topological Spaces

We consider pairs (X, i_X) where X is a topological space and i_X is an involution on X, i.e. a *continuous* self-map such that $i_X^2 = \mathrm{id}_X$. A map between two such pairs $(X, i_X), (Y, i_Y)$ is a continuous map $f : X \to Y$ such that

$$f \circ i_X = i_Y \circ f.$$

f is then called *equivariant* with respect to the involutions on X and Y. When there is no danger of confusion we shall omit the indices X and Y of the involution. We write $f : (X, i) \multimap (Y, i)$ to indicate that f is continuous and equivariant.

In principle every topological space X could be considered as a pair (X, i) by taking $i = \mathrm{id}$. But we shall see in a minute that only those pairs (X, i) are of interest here where the involution i is *fixpoint-free*, i.e. $i(x) \neq x$ for every $x \in X$.

3.1 Examples.

(1) $X = \mathbf{R}^n$ or $X = S^{n-1} = \{x \in \mathbf{R}^n : \|x\| = 1\}$, $i(x) = -x$ for $x = (x_1, \ldots, x_n)$.
   i is called the *antipodal* map.

(2) $X = \mathbf{C}^n$, $i(x_1, \ldots, x_n) = (\bar{x}_1, \ldots, \bar{x}_n)$ where $x_j \to \bar{x}_j$ is complex conjugation. Note that $\mathbf{R}^n \subset \mathbf{C}^n$ is exactly the set of fixpoints of this involution.

(3) Let $\mathfrak{a} < \mathbf{R}[X_1, \ldots, X_n]$ be an ideal. Let $X = V_{\mathbf{C}}(\mathfrak{a}) \subseteq \mathbf{C}^n$ be the affine algebraic set defined by $\mathfrak{a}$, i.e.

$$X = \{(x_1, \ldots, x_n) \in \mathbf{C}^n : f(x_1, \ldots, x_n) = 0 \text{ for every } f \in \mathfrak{a}\}.$$

X inherits the topology of $\mathbf{C}^n$. Since X is defined over $\mathbf{R}$ the complex conjugation i on $\mathbf{C}^n$ takes X to X. Hence (X, i) is a topological space with involution. (X, i) is fixpoint-free if and only if the ideal $\mathfrak{a}$ has no real zeros, say $\mathfrak{a} = (1 + X_1^2 + \ldots + X_n^2)$.

We are now able to define the level of a space (X, i) by comparing it with spheres and their antipodal involution $-$.

3.2 Definition. The number

$$s(X, i) = \min\{n : \text{there exists } f : (X, i) \multimap (S^{n-1}, -)\}$$

is called the *level of* (X, i). If there is no $n \in \mathbf{N}$ with this property we put $s(X, i) = \infty$.

3.3 Examples.

(1) If i has a fixpoint $x \in X$ then $s(X, i) = \infty$ since otherwise $f(x) \in S^{n-1}$ would have to be fixed under the antipodal map $-$.

(2) If there is a map $g : (X, i) \longrightarrow (Y, i)$ then clearly $s(X, i) \leq s(Y, i)$: compose g with a map $f : (Y, i) \longrightarrow (S^{n-1}, -)$ in the case $s(Y, i) = n < \infty$.

(3) If there are maps $g : (X, i) \longrightarrow (Y, i)$ and $h : (Y, i) \longrightarrow (X, i)$ in both directions then $s(X, i) = s(Y, i)$. This applies in particular if (X, i) and (Y, i) are homotopy-equivalent as topological spaces with involution. In this case we have in addition $h \circ g \sim \mathrm{id}_X$, $g \circ h \sim \mathrm{id}_Y$.

The most important result of this section is

3.4 Theorem. $s(S^{n-1}, -) = n$ for all $n \in \mathbf{N}$.

This is just a disguised form of the theorem of Borsuk–Ulam from topology which tells us that there is no map $(S^{n-1}, -) \longrightarrow (S^{m-1}, -)$ for $m < n$. We shall give a complete proof of this theorem in the next chapter.

Further examples of interest are given in the next proposition.

3.5 Proposition.

(1) Let $U \subseteq \mathbf{R}^n$, let $X = (U \times U) \setminus \Delta \subseteq \mathbf{R}^{2n}$ where $\Delta = \{(u, u) : u \in U\}$ is the diagonal of $U \times U$. Suppose $X \neq \emptyset$. Then X has the natural fixpoint-free involution $i : x = (u, v) \mapsto i(x) = (v, u)$ for $u, v \in U, u \neq v$. We get $s(X, i) \leq n$.

(2) Let (X, i) be a space with fixpoint-free involution such that X is a topological subspace of $\mathbf{R}^n$. Then $s(X, i) \leq n$.

(3) Let (X, i) be a subspace of $(\mathbf{C}^n, i)$ where i is the complex conjugation on $\mathbf{C}^n$. Suppose (X, i) fixpoint-free. Then $s(X, i) \leq n$.

PROOF. In all three cases it is easy to construct an equivariant map f into $(S^{n-1}, -)$:

(1) Define $f(x) = \frac{u-v}{\|u-v\|}$ for $x = (u, v) \in X$ where $\|u - v\|$ is the norm (length) of the vector $0 \neq u - v \in \mathbf{R}^n$.

(2) Define $f(x) = \frac{x-i(x)}{\|x-i(x)\|}$.

(3) Define $f(x) = \frac{\operatorname{Im} x}{\|\operatorname{Im} x\|}$ where $\operatorname{Im} x$ is the "imaginary part" of the point $x \in X \subseteq \mathbf{C}^n$. (For an arbitrary vector $z = (z_1, \ldots, z_n) \in \mathbf{C}^n$ with $z_j = x_j + iy_j$, $i = \sqrt{-1}$, x_j and $y_j \in \mathbf{R}$ we put $\operatorname{Im} z = y = (y_1, \ldots, y_n) \in \mathbf{R}^n$.) Note that $\operatorname{Im} x \neq 0$ for $x \in X$ since we assumed (X, i) fixpoint-free.

We mention a few other cases of topological spaces with involution for which the level can be computed or at least estimated from below and above. Dai and Lam [1984] study the space $V_{n,m}$ of orthogonal m-frames in $\mathbf{R}^n$ ($m \leq n$), the so-called Stiefel manifolds. There are two basic fixpoint-free involutions:

$$\delta \;:\; (v_1,\ldots,v_m) \mapsto (-v_1,\ldots,-v_m),$$
$$\varepsilon \;:\; (v_1,\ldots,v_m) \mapsto (v_1,\ldots,v_{m-1},-v_m) \;\text{ on } V_{n,m}.$$

It turns out that in some cases the level depends not only on the space $V_{n,m}$ but also on the involution, for instance one has $s(V_{n,2},\delta) = n$ for all $n \geq 2$ whereas

$$s(V_{n,2},\varepsilon) = \begin{cases} n - 1 & \text{for } n = 2,4,8 \\ n & \text{otherwise.} \end{cases}$$

Further interesting examples are the projective spaces of odd dimension over $\mathbf{R}$ or $\mathbf{C}$. We study them in some detail in Chapter 10.

The topological level of spaces (X,i) and the algebraic level of commutative rings are intimately related. This will be expressed in the next theorem. It is due to the observation that with any space (X,i) we get an $\mathbf{R}$-algebra of complex-valued functions on X as follows: Consider the set of equivariant continuous maps $f : (X,i) \multimap (\mathbf{C},\bar{\;})$. This set contains the *constant real-valued* functions on X. Furthermore two such functions f_1, f_2 can be added and multiplied pointwise:

$$(f_1 \overset{+}{\cdot} f_2)(x) = f_1(x) \overset{+}{\cdot} f_2(x)$$

for all $x \in X$. It is easily checked that all axioms for a commutative $\mathbf{R}$-algebra are satisfied for this set of functions.

Notation.
$$A_{X,i} := \{f : (X,i) \multimap (\mathbf{C},\bar{\;})\}$$
is called the $\mathbf{R}$-algebra of "functions on (X,i)".

3.6 Example. Let $g : (X,i) \multimap (Y,i)$ be a map between two topological spaces with involution. Then g induces an $\mathbf{R}$-algebra homomorphism $g^* : A_{Y,i} \to A_{X,i}$ as follows: For $f \in A_{Y,i}$ put $g^*(f) = f \circ g \in A_{X,i}$. g^* is unitary, i.e. $g^*(1) = 1$ where 1 denotes the constant function of value 1 on Y or X.

Note. In technical terms the correspondence $(X,i) \rightsquigarrow A_{X,i}$, $g \rightsquigarrow g^*$ is a contravariant functor from the category of topological spaces with involution to the category of (unitary) commutative $\mathbf{R}$-algebras.

We can now state and prove the second main theorem of this section.

3.7 Theorem. For every space (X, i) we have

$$s(X, i) = s(A_{X,i}).$$

PROOF. It will turn out that the proof is tricky, but completely elementary! Note that the letter i appears in two meanings during the proof, as an involution and as the imaginary unit $\sqrt{-1}$. This should not cause any confusion.

(1) Let $A(n) = A_{S^{n-1},\_}$ be the **R**-algebra of functions on S^{n-1}, the $(n-1)$-dimensional sphere. We show $s(A(n)) \le n$.

Let $x = (x_1, \ldots, x_n) \in S^{n-1}$ where $x_j \in \mathbf{R}$, $x_1^2 + \ldots + x_n^2 = 1$. Consider the functions

$$f_j(x) = ix_j \quad (j = 1, \ldots, n)$$

on S^{n-1} where $i = \sqrt{-1}$ is the imaginary unit. Then it is obvious that $f_j \in A(n)$ and

$$(f_1^2 + \ldots + f_n^2)(x) = -1(x_1^2 + \ldots + x_n^2) = -1.$$

Hence $s(A(n)) \le n = s(S^{n-1}, -)$ (see 3.4).

(2) For every space (X, i) we have

$$s(A_{X,i}) \le s(X, i).$$

For this we may suppose without loss of generality that $s(X, i) = n$ is finite. Then we have a map

$$g : (X, i) \dashrightarrow (S^{n-1}, -).$$

Let $g^* : A(n) \to A_{X,i}$ be the homomorphism induced by g and apply it to the equation

$$f_1^2 + \ldots + f_n^2 = -1$$

from part (1). This yields

$$g^*(f_1)^2 + \ldots + g^*(f_n)^2 = -1$$

in $A_{X,i}$. Hence $s(A_{X,i}) \le n = s(X, i)$.

(3) For every space (X, i) we have

$$s(X, i) \le s(A_{X,i}).$$

For this we may suppose that $s(A_{X,i}) = n$ is finite. Then we have functions $f_j \in A_{X,i}$ with

$$f_1^2 + \ldots + f_n^2 = -1.$$

Put $f_j = p_j + iq_j$ $(j = 1, \dots, n)$ where p_j, q_j are real-valued (continuous) functions on X. Assume for a moment that the functions $q_1, \dots, q_n$ have a common zero $x \in X$. Then we would get

$$p_1(x)^2 + \dots + p_n(x)^2 = -1 \quad \text{in } \mathbf{R},$$

a contradiction. Hence $q_1, \dots, q_n$ do not vanish simultaneously on X and define a continuous map

$$q : X \to S^{n-1}, \quad q(x) = \frac{(q_1(x), \dots, q_n(x))}{\sqrt{q_1^2(x) + \dots + q_n^2(x)}}.$$

The property $f_j(i(x)) = \overline{f_j(x)} = p_j(x) - iq_j(x)$ for the functions $f_j \in A_{X,i}$ implies $q_j(i(x)) = -q_j(x)$ and shows that the map q is equivariant, i.e. $q : (X, i) \multimap (S^{n-1}, -)$. Hence

$$s(X, i) \leq n = s(A_{X,i}).$$

The proof of 3.7 is finished. It implies in particular $s(A(n)) = s(S^{n-1}, -) = n$ for every $n \in \mathbf{N}$. This was necessary for the proof of Theorem 2.3.

Chapter 4

Hilbert's Homogeneous Nullstellensatz for p-fields and Applications to Topology

§1. The Nullstellensatz

In this section we use without proofs some well-known statements about noetherian rings and modules. For the convenience of the reader I repeat the most important facts:

(1) A module M over a commutative ring R is called noetherian if it satisfies (one of) the following three equivalent conditions:

    – every submodule is finitely generated,

    – every ascending chain of submodules becomes stationary,

    – every non-empty family of submodules contains a maximal submodule.

(2) A commutative ring R is called noetherian if R is a noetherian module over itself, or equivalently, if every ideal of R is finitely generated.

(3) Submodules and quotient modules of noetherian modules are noetherian.

(4) A module M over a noetherian ring R is noetherian if and only if M is finitely generated as an R-module.

(5) The polynomial ring $R = K[X_0, \ldots, X_n]$ over a field K is noetherian (for every $n \in \mathbf{N_0}$).
(This result is Hilbert's Basis Theorem.)

For more details and proofs see e.g. [Lang 1965, Chapter VI].

The following arrangement of lemmas and the elementary and reasonably short proof of the nullstellensatz are due to H.J. Fendrich (Mainz). They use the Hilbert polynomial but not the theorem of Bezout or results from intersection theory.

1.1 Definition. A polynomial $P(z) \in \mathbf{Q}[z]$ of degree $d \geq 0$ is called a *numerical polynomial* if

$$P(z) = c_0 \binom{z}{d} + c_1 \binom{z}{d-1} + \ldots + c_d$$

where $c_i \in \mathbf{Z}, c_0 \neq 0$, and $\binom{z}{d} = \frac{1}{d!} z(z-1) \ldots (z - d + 1)$.

For $d < 0$ we put $P(z) = 0$. Clearly numerical polynomials have the property $P(i) \in \mathbf{Z}$ for all $i \in \mathbf{Z}$. In the following $i \gg 0$ means "$i \in \mathbf{N}$ and big enough", i.e. $i \geq i_0$ for some $i_0 \in \mathbf{Z}$ depending on the given functions, coefficients etc. but not on the variable z.

1.2 Lemma. Suppose f is a function on $\mathbf{N}_0$ taking values in $\mathbf{Q}$ such that $f(i) \in \mathbf{Z}$ for $i \gg 0$ and

$$(\Delta f)(i) := f(i+1) - f(i) = Q(i) \quad \text{for } i \gg 0$$

where Q is a numerical polynomial of degree $d \geq 0$. Then there is a (unique) numerical polynomial P of degree $d + 1$ with

$$f(i) = P(i) \quad \text{for } i \gg 0.$$

PROOF. Let $Q(z) = c_0 \binom{z}{d} + \ldots + c_d$, put

$$P(z) = c_0 \binom{z}{d+1} + \ldots + c_d \binom{z}{1} + c_{d+1}$$

where $c_{d+1} \in \mathbf{Z}$ has to be determined. Since

$$\binom{z+1}{r+1} - \binom{z}{r+1} = \binom{z}{r} \quad \text{for } r \in \mathbf{N}_0$$

we have $(\Delta P)(z) = Q(z)$ identically in z. Our assumption on f implies $(\Delta(f - P))(i) = 0$ for $i \gg 0$, i.e. $f(i) - P(i)$ is constant for big i. For a suitable choice of c_{d+1} we get $f(i) - P(i) = 0$ for $i \gg 0$.

From now on we consider polynomial rings $R = K[X_0, \ldots, X_n]$ in finitely many indeterminates $X_0, \ldots, X_n$ ($n \geq 0$) over a fixed field K. We allow the case $n = -1$ where by definition $R = K$. For the moment n is fixed but later we will use induction on n. R is a graded ring, i.e. as a K-module we have

$$R = \bigoplus_{d \geq 0} R_d$$

where R_d consists of the homogeneous polynomials of exact total degree d and the zero element 0. R_d is a K-vector-space of dimension $\binom{d+n}{n}$. Similarly, an R-module M is called *graded* if

$$M = \bigoplus_{i \geq 0} M_i$$

with K-vector-spaces M_i such that $R_d M_i \subseteq M_{d+i}$ for all i and d. For $f \in R_d$ multiplication by f then gives rise to exact sequences:

$$(*) \qquad 0 \longrightarrow {}_fM_i \longrightarrow M_i \xrightarrow{f} M_{i+d} \longrightarrow M_{i+d}/fM_i \longrightarrow 0 \quad (i \geq 0).$$

Here ${}_fM_i := \{m \in M_i | fm = 0\}$ denotes the kernel of the multiplication by f when restricted to M_i. Note that the quotient module $M/fM \cong M_0 \oplus \dots \oplus M_{d-1} \bigoplus_{i \geq 0}(M_{i+d}/fM_i)$ is also graded.

1.3 Theorem. (Hilbert–Serre). For every finitely generated graded R-module M the submodules M_i are finite-dimensional K-vector-spaces and there is a unique numerical polynomial $P = H_M$ such that

$$\dim M_i = P(i) \quad \text{for } i \gg 0.$$

H_M is called the "Hilbert polynomial" of M.

PROOF. We use double induction on n and i.

(1) Let $n = -1$, i.e. $R = K$. Then M itself is a finite-dimensional vector-space. This implies $M_i = 0$ for $i \gg 0$. The theorem is satisfied with the (unique) numerical polynomial $P = 0$. (Without loss of generality one may assume $R = R_0, M = M_0$ in this trivial case.)

(2) Suppose now $R = K[X_0, \dots, X_n]$ with $n \geq 0$ and assume by induction that the theorem is true for all finitely generated graded modules over the "smaller" polynomial ring $R' = K[X_0, \dots, X_{n-1}]$. Put $X = X_n$. If M is the given R-module then the R-modules ${}_XM$ and M/XM are annihilated by (multiplication with) X, hence can be considered as finitely generated graded R'-modules. Therefore we have numerical polynomials $H_{{}_XM}$ and $H_{M/XM}$ such that

$$\begin{aligned} \dim({}_XM_i) &= H_{{}_XM}(i), \\ \dim(M_{i+1}/XM_i) &= \dim(M/XM)_{i+1} = H_{M/XM}(i+1). \end{aligned}$$

for $i \gg 0$. We now consider $(*)$ for $f = X, d = 1$ and use induction on i to show $\dim M_i < \infty$ and to compute $\dim M_i$. For $i = -1$ we have $M_{-1} = 0, \dim M_{-1} = 0$. Suppose we have already shown $\dim M_i < \infty$ for some $i \geq -1$. Then $(*)$ implies

$$\dim M_{i+1} - \dim M_i = \dim(M_{i+1}/XM_i) - \dim({}_XM_i)$$

for all $i \geq -1$. Here the dimensions on the right hand side are finite by induction on n since the modules are actually R'-modules, and $\dim M_i$ is finite by our present assumption. Thus $\dim M_{i+1}$ is finite, too. For large i the above equations give in addition

$$\dim M_{i+1} - \dim M_i = H_{M/XM}(i+1) - H_{{}_XM}(i) =: Q(i)$$

where Q is a numerical polynomial which is uniquely determined by M and $X = X_n$ but clearly depends only on M since this is true for the left hand side. We can now apply Lemma 1.2 and find

$$\dim M_i = P(i) \quad \text{for } i \gg 0$$

for some (unique) numerical polynomial $P = H_M$. Since we have already shown $\dim M_i < \infty$ for all i this proves the theorem.

In the following definition we exclude the trivial case $H_M = 0$, or equivalently $\dim_K M < \infty$.

1.4 Definition. For a finitely generated graded R-module M with Hilbert polynomial

$$H_M(z) = c_0 \binom{z}{d} + \ldots + c_d, \ c_0 \neq 0,$$

we call $\delta_M := \deg H_M = d$ the *degree* of M, and we call $\mu_M := c_0$ the *multiplicity* of M. Then $H_M(z) = \frac{\mu_M}{d!} z^d +$ lower terms (of degree $\leq d - 1$). (It is clear that $c_0 \neq 0$ implies $c_0 > 0$.)

1.5 Example. For $M = R = K[X_0, \ldots, X_n]$ we have

$$H_M(z) = \binom{z + n}{n}, \quad \text{i.e. } \delta_M = n, \mu_M = 1.$$

1.6 Definition. An R-module M is called *integral* if for any $f \in R$ we have $fM = 0$ or $_fM = 0$.

1.7 Example. Let P be a proper ideal in a commutative ring R. Then we have

$$M = R/P \text{ integral} \Leftrightarrow \ _fM = 0 \text{ for all } f \notin P \Leftrightarrow$$

$$P \text{ is a prime ideal} \Leftrightarrow M \text{ is an integral domain.}$$

In the following lemmas $R = K[X_0, \ldots, X_n]$ is the polynomial ring over a field K (with $n \geq 0$) with the natural grading $R = \oplus_{d \geq 0} R_d$, M is a finitely generated graded R-module of degree ≥ 0.

1.8 Lemma. Suppose M integral, $d > 0$, $0 \neq f \in R_d$ and $fM \neq 0$. Then

$$\delta_{M/fM} = \delta_M - 1, \ \mu_{M/fM} = d\mu_M.$$

PROOF. Consider $(*)$ for $i - d$ instead of i (where $i \geq d$) and use $_fM = 0$. Then we get for $i \gg 0$

$$H_{M/fM}(i) \quad = \quad H_M(i) - H_M(i - d)$$

$$= \frac{\mu}{\delta!}(i^\delta - (i - d)^\delta) + \text{ lower terms of similar shape}$$

$$= \frac{\mu \delta d}{\delta!} i^{\delta - 1} + \dots$$

$$= \frac{\mu d}{(\delta - 1)!} i^{\delta - 1} + \dots$$

where $\delta = \delta_M$, $\mu = \mu_M$. The lemma follows.

1.9 Lemma. Suppose p is a prime number with $p \nmid \mu_M$. Then M has a homogeneous submodule S such that the factor module $N = M/S$ satisfies the conditions

$$(1) \ \delta_N = \delta_M, \quad (2) \ p \nmid \mu_N, \quad (3) \ N \text{ is integral.}$$

PROOF. A submodule S of M is called homogeneous if it is generated by homogeneous elements. Then

$$S = \bigoplus_{i \geq 0} S_i \quad \text{with } S_i = S \cap M_i, \ M/S = \bigoplus_{i \geq 0} (M_i/S_i),$$

and both are graded R-modules. Since M is a noetherian R-module and since properties (1) and (2) are satisfied for $S = 0$ there exists a *maximal homogeneous* submodule S such that (1) and (2) are satisfied for $N = M/S$. We want to show that (3) follows from the maximality of S.

Suppose $f \in R$, $fN \neq 0$, and w.l.o.g. $f \in R_d$ ($d \geq 0$). (Otherwise decompose f into its homogeneous components.)

$(*)$ leads to the exact sequence

$$0 \longrightarrow N_{i-d}/{}_f N_{i-d} \longrightarrow N_i \longrightarrow (N/fN)_i \longrightarrow 0$$

for $i \geq d$ which implies

$$H_N(i) = H_{N/fN}(i) + H_{N/{}_f N}(i - d) \quad \text{for } i \gg 0.$$

Since $fN \neq 0$ we have $fM \not\subseteq S$.

This shows that the module $N/fN \cong M/(S + fM)$ must violate one of the properties (1) and (2). Hence either $\delta_{N/fN} < \delta_N = \delta_M$ or $\delta_{N/fN} = \delta_M$ and $p | \mu_{N/fN}$. In both cases the equation for $H_{N/{}_f N}$ gives $\delta_{N/{}_f N} = \delta_M$ and $p \nmid \mu_{N/{}_f N}$, i.e. (1) and (2) hold for $N/{}_f N = N'$. N' is of the form $N' = M/S'$ with a homogeneous submodule $S' \supseteq S$. By the maximality of S we must have $S' = S$ and ${}_f N = 0$. This shows that N is integral, i.e. N satisfies (1), (2) and (3).

1.10 Lemma. Let $P \subset R$ be a homogeneous prime ideal such that the integral R-module $M = R/P$ satisfies

$$\delta_M = 0, \quad \mu_M = c \neq 0.$$

Then:

(1) $XM \neq 0$ for some $X = X_j \in \{X_0, \ldots, X_n\}$.

(2) $L = M/(X - 1)M$ is a field, $[L : K] = c$.

PROOF.

(1) $X_iM = 0$ for all $i = 0, \ldots, n$ would imply $X_i \in P$ for all i, i.e. $M = K, H_M = 0$ contradicting $H_M = c$.

(2) Fix $X = X_j$ with $XM \neq 0$. Then $_XM = 0$ since M is integral. This shows that $M \xrightarrow{X} M$ is injective. For large i we have $\dim M_i = \dim M_{i+1} = c$. Therefore $M_i \xrightarrow{X} M_{i+1}$ is bijective for $i \gg 0$.

We now consider the (non-homogeneous!) submodule $(X - 1)M$ of M and its factor module $L = M/(X - 1)M$. Let π denote the natural epimorphism $M \to L$. By definition of L we have

$$X\pi(m) = \pi(Xm) = \pi(m)$$

for every $m \in M$, i.e. multiplication by $X \in R$ induces the identity on L, and for $m = \sum_{i=0}^{k} m_i$ with $m_i \in M_i$ we get $\pi(m) = \pi(\sum_{i=0}^{k} X^{k-i} m_i) = \pi(m')$ with $m' \in M_k$. This shows $L = \pi(M) = \pi(M_k) = \pi(M_{k+1}) = \ldots$ for $k \gg 0$. On the other hand a homogeneous element $m' \in M_k$ cannot be of the form $m' = (X - 1)m''$ unless $m' = 0$, since $X - 1$ is not homogeneous. Therefore $\pi|M_k$ is injective, and $\dim L = \dim M_k = c$ for $k \gg 0$. We now use that $M = R/P$ is not only an R-module but also an integral domain. Hence $L = \pi(M)$ is a commutative ring and a K-vector-space of dimension $c > 0$. It is easily seen that L is an integral domain: Suppose $\ell, \ell' \in L, \ell\ell' = 0$. By the above let $\ell = \pi(m), \ell' = \pi(m')$ with $m, m' \in M_k$. Then $0 = \ell\ell' = \pi(mm')$ with $mm' \in M_{2k}$. $\pi|M_{2k}$ is injective, hence $mm' = 0$, hence $m = 0$ or $m' = 0$ since M is integral. This gives $\ell = 0$ or $\ell' = 0$. Finally, it is a well-known fact that a finite-dimensional integral domain L over a field K is a field.

1.11 Definition. Let p denote a prime number. A field K is called a *p-field* if $[L : K]$ is a power of p for every finite field extension L/K.

1.12 Examples.

(1) If K is algebraically closed then $[L : K] = 1$ for every finite field extension L/K. Hence K is a p-field for all primes p. Conversely, if K is a p_1-field and a p_2-field for two different primes p_1, p_2 then K must be algebraically closed.

(2) **R** is a 2-field.

We come to the main theorem of this section.

1.13 Theorem. Let K be a p-field, $n \geq 1, R = K[X_0, \ldots, X_n]$. Let $f_1, \ldots, f_n \in R$ be homogeneous polynomials of degrees $d_1, \ldots, d_n$ where $p \nmid d_i$ $(i = 1, \ldots, n)$. Then there exists a common zero $0 \neq a = (a_0, \ldots, a_n) \in K^{n+1}$ of the system of equations

$$f_1 = \ldots = f_n = 0.$$

PROOF.

(1) We start with $M_0 = R$ and construct a sequence of integral (finitely generated, graded) R-modules

$$M_i = R/P_i \quad (i = 1, \ldots, n)$$

such that $f_1, \ldots, f_i \in P_i, M_i$ has degree $\delta_i = n - i$ and multiplicity $c_i \neq 0$ with $p \nmid c_i$. Actually the P_i will be homogeneous prime ideals, hence the M_i are integral domains. For $i = 0$ the conditions are satisfied with $P_0 = 0$ since $\delta_{M_0} = \delta_R = n, \mu_{M_0} = \mu_R = 1$. Inductively suppose we have already defined M_i for some $i \geq 0$. We apply Lemmas 1.8 and 1.9 to the pair (M_i, f_{i+1}) in order to find M_{i+1}:

a) If $f_{i+1} \notin P_i$ then $f_{i+1} M_i \neq 0$, and for

$$N_i := M_i / f_{i+1} M_i \cong R/(P_i + f_{i+1} R)$$

we have

$$\delta_{N_i} = \delta_i - 1 = n - (i + 1), \quad \mu_{N_i} = d_{i+1} \cdot c_i, \quad p \nmid d_{i+1} c_i$$

by Lemma 1.8.

b) If $f_{i+1} \in P_i$ we *replace* f_{i+1} by a suitable variable $X_{j(i+1)}$ such that $X_{j(i+1)} \notin P_i$.

Such a variable must exist: See part (1) of the proof of Lemma 1.10 which works for every module M of degree $\delta_m \geq 0$. As $X_{j(i+1)}$ has degree 1 we get $N_i := M_i / X_{j(i+1)} M_i, \delta_{N_i} = n - (i+1), \mu_{n_i} = c_i$ in this (exceptional) case.

c) N_i need not be integral but by Lemma 1.9 N_i has an integral factor module

$$M_{i+1} := N_i / S_i \cong R/P_{i+1}$$

with $\delta_{i+1} = \delta_{N_i} = n - (i + 1)$ and $p \nmid \mu_{M_{i+1}} =: c_{i+1}$. Clearly P_{i+1} is a homogeneous prime ideal and $f_1, \ldots, f_{i+1} \in P_{i+1}$.

(2) It remains to apply Lemma 1.10 to the last module M_n in our chain. We get a field

$$L = M_n/(X_j - 1)M_n \cong R/Q$$

with $[L : K] = c_n$, $p \nmid c_n$. Here Q is some non-homogeneous ideal of R. Since K is a p-field we must have $c_n = 1, L = K$. By construction we have $f_1, \ldots, f_n \in Q$, $X_j \equiv 1 \bmod Q$. Let a_i be the image of X_i in $L = K$ $(i = 0, \ldots, n)$. Then $a = (a_0, \ldots, a_n) \neq 0$ since $a_j = 1$, and

$$f_1(a) = \ldots = f_n(a) = 0.$$

This proves the theorem.

Note. Though our proof of 1.13 is purely algebraic it is not difficult to see the main geometric ideas behind it: To every homogeneous ideal $I \subset R$ there corresponds a projective algebraic set $V_I = \{a \in P^n(\bar{K}) : f(a) = 0$ for all $f \in I\}$ and its set $V_I(K)$ of K-rational points. If $I = (f_1, \ldots, f_n)$ then V_I is the intersection of n hypersurfaces $H_i = \{a \in P^n(\bar{K}) : f_i(a) = 0\}$. In a "generic situation" the intersection of $m \leq n$ hypersurfaces is an algebraic variety of dimension $n - m$. For $m = n$ we then get a finite set V_I and the general theorem of Bezout tells us that V_I has cardinality $c = d_1 \ldots d_n$ where $d_i = \deg f_i$. By a further argument one can show that at least one of the c points of V_I over $\bar{K}$ (the algebraic closure of K) must be K-rational provided K is a p-field with $p \nmid c$.

But the details of most proofs in the spirit of algebraic geometry are quite intricate because one has to decompose algebraic sets into irreducible components (varieties), study the dimension of varieties, define intersections and intersection-multiplicities, derive a Bezout theorem and (in the case where K is not algebraically closed) treat rationality and inseparability questions. In our proof I (resp. V_I) is replaced by the module (resp. ring) $M = R/I$, Lemma 1.8 corresponds to an intersection of V_I with a hypersurface, Lemma 1.9 corresponds to replacing an algebraic set by a suitable irreducible component, and Lemma 1.10 corresponds to the case $\dim V_I = 0$. In all three lemmas the behaviour of the multiplicity μ_M is carefully watched. M "integral" corresponds to "I prime ideal" or, equivalently, to "V_I irreducible".

Historical Note.

(1) For $K = \bar{K}$ algebraically closed Theorem 1.13 is implicitly contained in Hilbert's work [Hilbert 1893₂] on invariant theory and the inhomogeneous nullstellensatz. Many 19th century mathematicians knew and used the result without having a strong proof since it seems to be "geometrically obvious".

(2) For $K = \bar{K}$ it is possible to derive the homogeneous nullstellensatz from the ordinary (inhomogeneous) nullstellensatz by a clever trick. This proof is given in [S, p.98].

(3) The first proofs of 1.13 in the case $K = \mathbf{R}, p = 2$ are from Borsuk, Hopf and Stiefel and are of topological nature (see [Hopf 1940]). The first purely algebraic proof is due to [Behrend 1940]. Another proof (for a real-closed field $K = R$) has been given by [Lang 1953].

(4) The first proof for arbitrary p-fields is due to [Terjanian 1972]. It uses a lot of local algebra, multiplicities, Bezout theorem etc. Another proof using henselizations and Bezout is due to Arason (see [Pfister 1979]). A proof of a more general theorem, based on intersection theory, is given in Chapter 13 of [Fulton 1984].

§2. The Borsuk–Ulam and Brouwer Theorems

In this section we apply the nullstellensatz for 2-fields in order to derive short and almost elementary proofs of the two famous theorems in the title.

2.1 Theorem. (Borsuk–Ulam for polynomial maps)
For $n \in \mathbf{N}$ let $q_1, \ldots, q_n \in \mathbf{R}[X_1, \ldots, X_{n+1}]$ be "odd" polynomials, i.e.

$$q_i(-X) = -q_i(X) \quad (i = 1, \ldots, n)$$

where $X = (X_1, \ldots, X_{n+1}), -X = (-X_1, \ldots, -X_{n+1})$. Then there exists a point $a = (a_1, \ldots, a_{n+1}) \in \mathbf{R}^{n+1}$ with $\sum_{i=1}^{n+1} a_i^2 = 1$ and $q_1(a) = \ldots = q_n(a) = 0$, i.e. $q_1, \ldots, q_n$ have a common zero on the unit sphere $S^n \subset \mathbf{R}^{n+1}$.

PROOF. We start by introducing an additional indeterminate X_0. Let $\tilde{q}_i = \tilde{q}_i(X_0, \ldots, X_{n+1})$ be the homogenized polynomial belonging to q_i, with $\deg \tilde{q}_i = \deg q_i = d_i$. Since the q_i contain only monomials of odd degree d_i is odd and X_0 occurs in $\tilde{q}_i$ in even powers only. Thus it is possible to replace

$$X_0^2 \quad \text{by} \quad X_1^2 + \ldots + X_{n+1}^2 \quad \text{in } \tilde{q}_i.$$

This leads to homogeneous polynomials $f_i(X_1, \ldots, X_{n+1})$ of odd degrees $d_i (i = 1, \ldots, n)$. Apply Theorem 1.13 with $K = \mathbf{R}, p = 2$. This shows that $f_1, \ldots, f_n$ have a common nontrivial zero $a = (a_1, \ldots, a_{n+1}) \in \mathbf{R}^{n+1}$. Since the f_i are homogeneous we can scale a by the factor $(\sum_i a_i^2)^{-1/2}$ in order to get $a \in S^n$. Going back to the original polynomials q_i we get

$$q_i(a) = \tilde{q}_i(1, a) = f_i(a) = 0 \quad (i = 1, \ldots, n).$$

2.2 Theorem. (Borsuk–Ulam for continuous maps) Let $f : S^n \to \mathbf{R}^n$ with $f = (f_1, \ldots, f_n)$ be an "odd" continuous map, i.e. $f(-x) = -f(x)$ for all $x \in S^n$. Then f vanishes for at least one point $a \in S^n$.

Corollary. There exists no continuous map $g : S^n \to S^{n-1}$ with $g(-x) = -g(x)$ for all $x \in S^n$.
(This is equivalent to Theorem 3.4 of Chapter 3 since $S^m \subseteq S^{n-1}$ for every $m \leq n - 1$.)

PROOF.

(1) Since S^n is compact we can apply the (Weierstraß part of the) theorem of Stone and Weierstraß (for a proof see e.g. [Burkill–Burkill 1970, chapter 5.6]): For any real $\varepsilon > 0$ there exist polynomials $p_i \in \mathbf{R}[X_1, \ldots, X_{n+1}]$ such that

$$(1) \qquad |f_i(x) - p_i(x)| < \varepsilon \quad \text{for all } x \in S^n \text{ and } i = 1, \ldots, n.$$

Putting $q_i(X) = \frac{1}{2}(p_i(X) - p_i(-X))$ one immediately derives that the p_i may be replaced by the odd polynomials q_i in (1) since $f_i(-x) = -f_i(x)$.

(2) *Assume* by contradiction that the given map $f : S^n \to \mathbf{R}^n$ has no zero. Then, since S^n is compact, there exists a $\delta > 0$ such that

$$(2) \qquad \max\{|f_1(x)|, \ldots, |f_n(x)|\} \geq \delta$$

for all $x \in S^n$. Choose $0 < \varepsilon < \delta$ in part (1). Then

$$\max\{|q_1(x)|, \ldots, |q_n(x)|\} \geq \delta - \varepsilon > 0$$

for all $x \in S^n$. This contradicts Theorem 2.1.

We shall now derive the Brouwer fixpoint theorem from Theorem 2.2. For this we use the following

Notation. $E^n = \{(x_1, \ldots, x_n) \in \mathbf{R}^n : \sum_1^n x_i^2 \leq 1\}$ is the n-ball,

$E_{\pm}^n = \{(x_1, \ldots, x_{n+1}) \in \mathbf{R}^{n+1} : \sum_1^{n+1} x_i^2 = 1, \pm x_{n+1} \geq 0\}$ is the upper, resp. lower, hemisphere of $S^n \subset \mathbf{R}^{n+1}$, $\partial E^n = S^{n-1} \cong E_+^n \cap E_-^n$ is the border of E^n, $\pi : \mathbf{R}^{n+1} \to \mathbf{R}^n$ is the projection map, given by $(x_1, \ldots, x_{n+1}) \mapsto (x_1, \ldots, x_n)$. Then we have

2.3 Proposition. There is no continuous map

$$r : E^n \to S^{n-1} \quad \text{with } r|_{S^{n-1}} = \text{id}$$

(such a map is called a "retract" in topology).

PROOF. Suppose that r exists. Define $f : S^n \to S^{n-1}$ by

$$f(x) = \begin{cases} r(-\pi(x)) & \text{for } x \in E^n_+, \\ -r(\pi(x)) & \text{for } x \in E^n_-. \end{cases}$$

f is well-defined: For $x \in E^n_+ \cap E^n_-$ we have $x_{n+1} = 0, \pi(x) = x$ and $r(-x) = -x = -r(x)$. f is clearly continuous. Finally, f is an antipodal map: For $x \in E^n_+$ we have

$$f(-x) = -r(\pi(-x)) = -r(-\pi(x)) = -f(x).$$

A similar equation holds for $x \in E^n_-$. Thus f contradicts the corollary to 2.2.

2.4 Fixpoint Theorem of Brouwer. Every continuous map $f : E^n \to E^n$ has a fixpoint $a \in E^n$ with $f(a) = a$.

PROOF. Suppose $f(x) - x \neq 0$ for all $x \in E^n$ and look at the figure below:

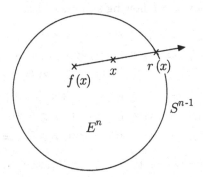

Define $r(x)$ to be the intersection of the arrow joining $f(x)$ and x (in this order) with $\partial E^n = S^{n-1}$. Then r is a continuous retract from E^n to S^{n-1}: contradiction.

Notes.

(1) Theorem 2.1 remains true if the field **R** of real numbers is replaced by an arbitrary real closed field R (for the definition see Chapter 6). Since the field $C = R(\sqrt{-1})$ is algebraically closed, R is a 2-field. Also every sum of squares in R is itself a square. Having these properties the proof of 2.1 goes through without change.

(2) On the other hand, 2.2, 2.3 and 2.4 cannot be generalized to a real closed field R which is not isomorphic to **R**, even for $n = 1$. Let's outline why: There are two types of real closed fields R:

a) R is *archimedean* over **Q**, i.e. to any $r \in R$ there exists an $n \in \mathbf{N}$ such that $-n < r < n$ in the order $<$ of R. Then the map

$$r \mapsto \sup\{q \in \mathbf{Q} : q < r\}$$

defines an order-isomorphic imbedding of R in **R**. Therefore we may assume $R \subset \mathbf{R}, R \neq \mathbf{R}$ in the archimedean case. An example is given by the field $R = \mathbf{R}_a$ of all real algebraic numbers.

b) R contains *infinitely large* elements $r \in R$ with $r > n$ for all $n \in \mathbf{N}$. Then $\varepsilon = \frac{1}{r}$ is an *infinitely small* element, i.e. $-q < \varepsilon < q$ for every $0 < q \in \mathbf{Q}$. An example is given by

$$R = \bigcup_{n \in \mathbf{N}} \mathbf{R}((t^{1/n})),$$

the field of all real Puiseux series in t. Here $r - t$ is a square for every positive $r \in \mathbf{R}$, since $(1 - \frac{t}{r})^{1/2} = \sum_{n=0}^{\infty} \binom{\frac{1}{2}}{n} \left(\frac{-t}{r}\right)^n \in R$. Hence $0 < t < r$ for every $0 < r \in \mathbf{R}$ which shows that t is infinitely small.

c) Consider now the following subsets of R:

$$
\begin{aligned}
E^1 &= \{r \in R : -1 \le r \le 1\}; \\
U_0 &= \{r \in R : |r| < r_0\} \quad \text{in case (a)}, \\
&\qquad \text{where } r_0 \in \mathbf{R}\backslash R,\ 0 < r_0 < 1; \\
U_0 &= \{\varepsilon \in R : \varepsilon \text{ infinitely small}\} \text{ in case (b)}; \\
U_1 &= \{r \in E^1 : r > 0,\ r \notin U_0\}; \\
U_{-1} &= \{r \in E^1 : r < 0,\ r \notin U_0\} = -U_1.
\end{aligned}
$$

Then it is easily seen that U_{-1}, U_0, U_1 are open in the order topology (also called the interval or euclidean or strong topology) of E^1 and that $E^1 = U_{-1} \cup U_0 \cup U_1$. Therefore E^1 is *not connected* and *not compact* in its topology. (These are the decisive properties which prevent carrying over the proofs for **R** to R.) This immediately gives the following continuous maps f, g contradicting the analogy of 2.3 and 2.4:

$$f : E^1 \to S^0 = \{1, -1\}, \quad f(U_0 \cup U_1) = 1, \quad f(U_{-1}) = -1,$$

$$g : E^1 \to E^1, \quad g(U_0 \cup U_1) = -1, \quad g(U_{-1}) = 1.$$

Finally, let $S^1 = \{(x, y) \in R^2 : x^2 + y^2 = 1\}$, $V_1 = \{(x, y) \in S^1 : \text{either } x \in U_1, \text{ or } x \in U_0 \text{ and } y > 0\}$, $V_{-1} = S^1 \backslash V_1 = -V_1$. Then S^1 is the disjoint union of the open subsets V_1, V_{-1} and

$$h : S^1 \to S^0, \quad h(V_1) = 1, \quad h(V_{-1}) = -1$$

is a continuous antipodal map contradicting the analogy of 2.2.

d) For a real closed field R of type (b) the approximation theorem of Weierstraß does not hold. It can be shown that there is no polynomial $g(x) \in R[x]$ which approximates $f(x) = |x|$ better than ε on the interval $E^1 = [-1, 1]$ (for every infinitely small $\varepsilon > 0$).

Historical Note. The fixpoint theorem was proved in [Brouwer 1912], the Borsuk–Ulam theorem was conjectured by Ulam and proved by [Borsuk 1933]. The first algebraic proof of Theorem 2.1 is probably the one by [Knebusch 1982]. The proof given here is in [Arason–Pfister 1982].

Chapter 5

Tsen–Lang Theory for C_i^p-fields

§1. The Main Theorems

The homogeneous nullstellensatz for an algebraically closed field C is the basis for a couple of similar results on "forms in many variables" which are valid in fields of finite transcendence degree over C. Similarly, starting with a p-field K instead of an algebraically closed field C and considering only forms of degrees relatively prime to p, these results can be generalized to C_i^p-fields which we are going to define now.

1.1 Definition. Fix $i \in \mathbb{N}_0$ and a prime p.

a) A field K is called a C_i-field if for any $r \in \mathbb{N}$ and any system $(f_1, \ldots, f_r)$ of forms

$$f_\varrho \in K[X_1, \ldots, X_n] \quad (\varrho = 1, \ldots, r)$$

of degree $d_\varrho = \deg f_\varrho > 0$ the following holds:

(*)   "$f_1, \ldots, f_r$ have a common nontrivial zero in K",

i.e. there is an $a \in K^n$, $a \neq 0$, with $f_1(a) = \ldots = f_r(a) = 0$, *provided* we have $n > d_1^i + \ldots + d_r^i$.

b) K is called a C_i^p-field if (*) holds under the additional assumption $p \nmid d_\varrho$ for $\varrho = 1, \ldots, r$.

1.2 Proposition. We have

(1) K is a C_i-field $\iff$ K is a C_i^p-field for all primes p.

(2) K is a C_0-field $\iff$ K is algebraically closed.

(3) K is a C_0^p-field $\iff$ K is a p-field.

PROOF.

(1) Trivially $C_i \implies C_i^p$ for every prime p. Conversely, if K is C_i^p for all primes p then given any system $(f_1, \ldots, f_r)$ of forms over K choose $p > \max(d_1, \ldots, d_r)$ to see that the equations $f_1 = \ldots = f_r = 0$ have a common nontrivial zero $a \in K^n$ whenever $n > \sum_1^r d_\varrho^i$.

(2) If K is C_0 then any form $a_d X^d + a_{d-1} X^{d-1} Y + \ldots + a_0 Y^d$ of degree $d > 0$ in two variables X, Y has a nontrivial zero (x, y) in K since $2 > d^0 = 1$. For $a_d \neq 0$ we must have $y \neq 0$. Hence any non-constant polynomial $a_d X^d + \ldots + a_0$ over K has a zero in K, i.e. K is algebraically closed. The converse statement is exactly the homogeneous nullstellensatz for K.

(3) The implication $\Longleftarrow$ is exactly the homogeneous nullstellensatz for p-fields.

Let us now assume that K is *not* a p-field. Then K has a field extension L of degree $p^r d$ with $r \geq 0$, $d > 1$ and $p \nmid d$.

a) L/K is separable: Replacing L by its normal closure we may assume that L/K is a Galois extension. Consider the fixed field L_0 of a Sylow p-subgroup of the Galois group. Then $[L_0 : K] = d > 1$, $p \nmid d$.

b) L/K is inseparable, char $K = p$. Then a) applies to the separable part L_s of L/K since $d \| [L_s : K]$.

c) L/K is inseparable, char $K = q \neq p$. Then K is not perfect (since it has an inseparable algebraic extension) which means that there is an element $a \in K$ which is not a q-th power in K. Then $[K(\sqrt[q]{a}) : K] = q > 1$, $p \nmid q$.

In any case we have constructed a new field extension M/K such that $[M : K] = d > 1$ with $p \nmid d$. We may assume that $M = K(\alpha)$ is a simple extension. The minimal polynomial of α, say $f(x) = x^d + a_{d-1} x^{d-1} + \ldots + a_0$, is irreducible over K. In particular $a_0 \neq 0$ and f has no zero in K. This implies that the form

$$X^d + a_{d-1} X^{d-1} Y + \ldots + a_0 Y^d$$

in $n = 2$ variables has no nontrivial zero in K though $2 > d^0 = 1$ and $p \nmid d$. Hence K is not a C_0^p-field. This completes the proof of (3).

1.3 Theorem. Let K be a C_i^p-field, let L be an algebraic extension field of K. Then L is a C_i^p-field.

PROOF. Let $r \in \mathbb{N}$, let $f_\varrho(X) \in L[X]$ be forms of degrees d_ϱ where $p \nmid d_\varrho$ ($\varrho = 1, \ldots, r$) and $X = (X_1, \ldots, X_n)$ with

$$n > d_1^i + \ldots + d_r^i.$$

We have to show that the system $(f_1, \ldots, f_r)$ has a common nontrivial zero in L.

The coefficients of $f_1, \ldots, f_r$ generate some finite algebraic extension field over K. Hence we may assume that L/K is finite, say $[L : K] = s < \infty$. Let $\omega_1, \ldots, \omega_s$ be a K-basis of L. Then any coefficient α of any f_ϱ has the form

$$\alpha = a_1 \omega_1 + \ldots + a_s \omega_s, \quad a_\sigma \in K.$$

Similarly, the indeterminates X_ν over L ($\nu = 1, \ldots, n$) are written in the form

$$X_\nu = X_{\nu 1} \omega_1 + \ldots + X_{\nu s} \omega_s$$

where the $X_{\nu\sigma}$ are considered as indeterminates over K.

Substituting the above expressions for α and X_ν into f_ϱ and reordering in terms of the basis $\omega_1, \ldots, \omega_s$ we get

$$f_\varrho(X_1, \ldots, X_n) = g_{\varrho 1}(X_{\nu\sigma})\omega_1 + \ldots + g_{\varrho s}(X_{\nu\sigma})\omega_s \quad (\varrho = 1, \ldots, r)$$

where $g_{\varrho\sigma}$ is a form of degree $d_{\varrho\sigma} = d_\varrho$ over K in the new variables $X_{11}, \ldots, X_{ns}$. We immediately see:
The r forms f_ϱ have a nontrivial common zero in L $\iff$ the rs forms $g_{\varrho\sigma}$ have a nontrivial common zero in K.
The number of new variables equals $s \cdot n$. From the assumptions we have

$$sn > s \sum_{\varrho=1}^{r} d_\varrho^i = \sum_{\varrho=1}^{r} \sum_{\sigma=1}^{s} d_{\varrho\sigma}^i, \quad p \nmid d_{\varrho\sigma}.$$

Since K is a C_i^p-field the new system $\{g_{\varrho\sigma}\}$ has a (nontrivial) zero in K, hence the old system $\{f_\varrho\}$ has a (nontrivial) zero in L.

1.4 Theorem. Let K be a C_i^p-field, let $L = K(Y)$ be the rational function field in one variable Y over K. Then L is a C_{i+1}^p-field.

PROOF. We imitate the substitution method of the last proof. Let $r \in \mathbf{N}$, let $f_\varrho(X) \in L[X] = L[X_1, \ldots, X_n]$ be forms of degrees d_ϱ where $p \nmid d_\varrho$, and suppose

$$n > d_1^{i+1} + \ldots + d_r^{i+1}.$$

We have to show that $\{f_1, \ldots, f_r\}$ have a common zero in L. By multiplying the coefficients $\alpha \in L = K(Y)$ of the forms f_ϱ by their common denominator we may suppose $\alpha \in K[Y]$, $f_\varrho \in K[Y; X_1, \ldots, X_n]$. Let t be an upper bound for the degrees of the coefficients $\alpha = a_0 + a_1 Y + \ldots + a_t Y^t$ ($a_\tau \in K$). We look for a solution of our system $\{f_1, \ldots, f_r\}$ in the ring $K[Y] \subset L$. Therefore we put

$$X_\nu = X_{\nu 0} + X_{\nu 1} Y + \ldots + X_{\nu s} Y^s \quad (\nu = 1, \ldots, n)$$

where the $X_{\nu\sigma}$ are indeterminates over K and where $s \in \mathbf{N}$ will be chosen later (large enough). We get

$$f_\varrho(X) = g_{\varrho 0}(X_{\nu\sigma}) + \ldots + g_{\varrho, sd_\varrho+t}(X_{\nu\sigma})Y^{sd_\varrho+t}$$

where $g_{\varrho\tau}$ is a form of degree d_ϱ over K ($\tau = 0, \ldots, sd_\varrho + t$). The number of new variables equals $(s+1)n$. The "degree sum" for the new forms $g_{\varrho\tau}$ equals

$$\sum_{\varrho=1}^{r}(sd_\varrho + t + 1)d_\varrho^i = s\sum d_\varrho^{i+1} + (t+1)\sum d_\varrho^i.$$

In order that the forms $g_{\varrho\tau}$ have a common nontrivial zero over the C_i^p-field K we need

$$(s+1)n > s\sum d_\varrho^{i+1} + (t+1)\sum d_\varrho^i,$$

i.e.

$$s(n - \sum d_\varrho^{i+1}) > (t+1)\sum d_\varrho^i - n.$$

Since $n > \sum d_\varrho^{i+1} \geq 0$ it is sufficient to take

$$s \geq (t+1)\sum d_\varrho^i.$$

Then the $g_{\varrho\tau}$ have a nontrivial zero over K which leads to a nontrivial zero over $K[Y]$ (of degree $\leq s$) for the original system $\{f_\varrho\}$.

1.5 Corollary (to 1.3 and 1.4). Let K be a C_i^p-field, let L be an extension field of finite transcendence degree $j \geq 0$ over K (L need not be finitely generated). Then L is a C_{i+j}^p-field.

PROOF. This is clear by repeated application of 1.4 and 1.3 since L is algebraic over $K(Y_1, \ldots, Y_j)$ where $Y_1, \ldots, Y_j$ is a transcendence basis of L/K.

1.6 Notes.

(1) The essential ideas go back to Chiungtze Tsen [1933 for $i = 1$, 1936 for arbitrary i] and Serge Lang [1952]. Tsen called C_1-fields "quasi-algebraically closed". It seems that his second paper (being published in a Chinese journal which finished after the publication of volume 2) got forgotten during the war. Serge Lang rediscovered the results in his thesis but refers only to Tsen's first paper.

(2) It should be pointed out that the usual definition of the property C_i (introduced by S. Lang) is slightly weaker because condition $(*)$ in Definition 1.1a) is only required for $r = 1$. It is not known whether $(*)$ for $r = 1$ implies $(*)$ for arbitrary r. On the other hand I am not aware of any counter-examples.

(3) There are a few cases where $(*)$ for $r = 1$ implies $(*)$ for any r under additional assumptions:

   a) (Nagata) This holds if $d_1 = \ldots = d_r = d$.

b) (Tsen) A normic form of level $i > 0$ is a form of degree $d > 1$ in
$n = d^i$ variables which has only the trivial zero over K. With this
terminology the following hold:

(i) If K admits a normic form of level i and *some* degree $d_0 > 1$
then $(*)$ for $r = 1$ implies $(*)$ for the case $d_1 = \ldots = d_r = d$
(all r and d). [This result is of course superseded by the result
of Nagata.]

(ii) If K admits normic forms of level i for *every* degree $d_0 > 1$ then
$(*)$ for $r = 1$ implies $(*)$ for any r.

The proofs use the so-called "Artin's trick" which consists of con-
structing normic forms of large degree and substituting the given
system $f_1, \ldots, f_r$ as often as possible into a normic form. For details
see the original papers of Tsen and Lang or the books of Greenberg
[1969] and Lorenz [1990].

(4) For $p = 2$ the C_i^2-fields are called "oddly C_i" in Lang's second paper
[Lang 1953]. As far as I know C_i^p-fields for $p > 2$ have not been studied
up to now.

§2. Some Related Results and Conjectures

Our next result was conjectured by E. Artin and proved by Chevalley [1936]:

2.1 Theorem. Finite fields are strongly C_1, i.e. for any $r \in \mathbb{N}$ and any
polynomials $f_1, \ldots, f_r$ of positive degrees $d_1, \ldots, d_r$ in n common variables
over a finite field $K = \mathbb{F}_q$ with $f_1(0) = \ldots = f_r(0) = 0$ there exists a nontrivial
zero $0 \neq a \in K^n$ of the system

$$f_1 = \ldots = f_r = 0$$

provided we have $n > d_1 + \ldots + d_r$.

Note that it is only supposed that the constant coefficients of $f_1, \ldots, f_r$
vanish. This is trivially the case if f_i is homogeneous of degree $d_i > 0$. In
particular every finite field is a C_1-field.

PROOF.

(1) Since $K^\bullet$ is a group of order $q-1$ we have $a^{q-1} = 1$ for all $a \in K^\bullet$, $a^q = a$
for all $a \in K$. This implies that K is contained in the zero set of the
polynomial $X^q - X$. Let now $f(X) = f(X_1, \ldots, X_n) \in K[X_1, \ldots, X_n]$

be a polynomial in n variables. We are interested in its values $f(a)$ for $a = (a_1, \ldots, a_n) \in K^n$. If $f(X) = X_1^{i_1} \ldots X_n^{i_n}$ is a monomial define

$$
j_\nu = \begin{cases} i_\nu & \text{for } i_\nu = 0, \\ \text{least } positive \text{ residue of } i_\nu \bmod (q-1) & \text{for } i_\nu > 0 \end{cases}
$$

$(\nu = 1, \ldots, n)$. Put $f_*(X) = X_1^{j_1} \ldots X_n^{j_n}$. By the above it is clear that $a^{i_\nu} = a^{j_\nu}$ for every $a \in K$, hence

$$
f(a) = f_*(a) \quad \text{for every } a \in K^n.
$$

Similarly, if f is an arbitrary polynomial we can apply the $*$-operation to every monomial which appears as a summand of f. This leads to a well-defined new polynomial f_* such that $f(a) = f_*(a)$ for every $a \in K^n$. f_* has the property: $\deg_{X_\nu}(f_*) \leq q-1$ for every $\nu \in \{1, \ldots, n\}$. f_* is called "the reduced polynomial of f".

(2) A reduced polynomial $g(X_1, \ldots, X_n) \in K[X_1, \ldots, X_n]$ which vanishes at every point $a \in K^n$ is identically zero. This results from an easy induction on n:

For $n = 0$ g is a constant and $g(0) = 0$, hence $g = 0$. For $n > 0$ write $g(X_1, \ldots, X_n) = \sum_{i=0}^{q-1} g_i(X_1, \ldots, X_{n-1}) X_n^i$. Fix $a' = (a_1, \ldots, a_{n-1}) \in K^{n-1}$. Then

$$
g(a', X_n) = \sum_{i=0}^{q-1} g_i(a') X_n^i \in K[X_n]
$$

is a polynomial of degree $< q$ which vanishes for the q different elements $a_n \in K$. Hence $g_i(a') = 0$ for $i = 0, \ldots, q-1$ and all $a' \in K^{n-1}$. By induction we get $g_i = 0$ (since the g_i are reduced), hence $g = 0$.

(3) $h(X_1, \ldots, X_n) = \prod_{\nu=1}^{n}(1 - X_\nu^{q-1})$ is the only reduced polynomial with the properties

$$
h(0) = 1, \ h(a) = 0 \quad \text{for } 0 \neq a \in K^n :
$$

For $a = (a_1, \ldots, a_n) \neq 0$ at least one factor in the product for $h(a)$ vanishes. This shows that h has the two properties. By (2) a reduced polynomial h is uniquely determined by its value set $\{h(a) : a \in K^n\}$.

(4) After these preliminary observations on polynomials over the finite field K we may now consider the given system $f_1, \ldots, f_r \in K[X_1, \ldots, X_n]$. Define

$$
f(X) = (1 - f_1(X)^{q-1}) \ldots (1 - f_r(X)^{q-1}).
$$

Then $\deg f = (q-1)(d_1 + \ldots + d_r) < (q-1)n = \deg h$. This implies $\deg f_* \leq \deg f < \deg h$, in particular $f_* \neq h$. From the assumption $f_1(0) = \ldots = f_r(0) = 0$ we get $f_*(0) = f(0) = 1 = h(0)$. Therefore

$f_* \neq h$ and (3) imply: There exists an $a \in K^n$, $a \neq 0$, such that $f(a) = f_*(a) \neq 0$. By the definition of f this implies $f_1(a) = \ldots = f_r(a) = 0$ because for, say, $f_1(a) = b_1 \neq 0$ we would get $f_1(a)^{q-1} = b_1^{q-1} = 1$ in K, hence $f(a) = 0$. This finishes the proof of Chevalley's theorem.

Without proof we mention

2.2 Theorem. Let K be a C_i^p-field, let $L = K((Y))$ be the field of formal Laurent series in one variable Y over K. Then L is a C_{i+1}^p-field.

For an algebraically closed or a finite field K the proof is already in the papers of Tsen and Lang quoted above, as well as in the book of Greenberg [1969]. For an arbitrary C_i-field K the first proof was given by [Greenberg 1966]. The main point is to show that the result for $L = K((Y))$ follows from Theorem 1.4 (i.e. the corresponding result for $K(Y)$) by an "approximation process": Cut off the coefficients $\alpha \in K((Y))$,

$$\alpha = \sum_{i=i_0}^{\infty} a_i Y^i, \quad a_i \in K,$$

of the forms f_ϱ at some power Y^t ($t \in \mathbb{N}$, large). Get $\alpha^{(t)} = \sum_{i=i_0}^{t} a_i Y^i$ and $f_\varrho^{(t)}$ (replace every α by $\alpha^{(t)}$ in f_ϱ). Then for every t the system $(f_1^{(t)}, \ldots, f_r^{(t)})$ has a nontrivial solution in $K[Y] \subset K[[Y]] \subset L$. Using the completeness of L it remains only (but this is the hard part) to show that this implies a nontrivial solution of the given system $(f_1, \ldots, f_r)$ in L. For a simplified proof of Greenberg's main theorem see [M. Kneser 1978]. Finally, the proof carries over verbatim to the more general case of the property C_i^p instead of C_i.

Theorems 1.3, 1.4, 2.1 and 2.2 are about all that can be said on the positive side about the C_i-properties. The rest is a lot of conjectures and of negative results.

2.3 Artin's Conjecture on $\mathbb{Q}^{ab}$. Let L be an (infinite) number field which contains all roots of unity, for instance the maximal cyclotomic extension of $\mathbb{Q}$ (which by a famous theorem of Kronecker–Weber coincides with the maximal abelian extension $\mathbb{Q}^{ab}$ of $\mathbb{Q}$). Is it true that L is a C_1-field?

Despite many efforts of number theorists this conjecture seems to be still open. The local counterpart of it has been proved by S. Lang in his thesis [1952]. It reads as follows:
Let K be a complete field under a discrete valuation with perfect residue field. Then the maximal unramified extension $L = K_{nr}$ of K is a C_1-field.

2.4 Lang's Conjecture. [Lang 1953] Let K be a field of finite transcendence degree i over a real closed field R. Suppose in addition that K is nonreal (i.e. -1 is a sum of squares in K). Is it true that K is a C_i-field?

Since R is a 2-field, it is clear by 1.5 that K is a C_i^2-field. In particular every form f of *odd* degree d in $n > d^i$ variables over K has a nontrivial zero in K. But what about a form f or a system $(f_1, \ldots, f_r)$ of even degree? For $i > 1$ not even the simplest case, namely *one quadratic form* in $n > 2^i$ variables, is known.

For $i = 1$ it is known that one quadratic form f in $n > 2$ variables over K is necessarily isotropic. This result is originally due to Witt [1934] and will be re-proved in the next chapter.

But even the cases of *one form of degree 4* or a *pair of quadratic forms* are open. In both cases $n = 5$ is the critical number of variables.

So we must admit that practically nothing is known in support of Lang's conjecture. But on the other hand, no one has found a counter-example.

2.5 Artin's Conjecture on $\mathbf{Q}_p$. Let p be a prime. Is the field $\mathbf{Q}_p$ of p-adic numbers a C_2-field?

(a) This conjecture is very natural since $\mathbf{Q}_p$ is complete with respect to a discrete valuation v_p with finite residue field $\mathbf{F}_p$. Hence $\mathbf{Q}_p$ has many similar properties to the power series field $\mathbf{F}_p((Y))$ which is C_2 by Theorems 2.1 and 2.2.

(b) Unfortunately, the conjecture is false. The first counter-example was found by [Terjanian 1966]. He constructs an anisotropic form of degree $d = 4$ in $n = 20$ variables over $\mathbf{Q}_2$. Soon after, he [Terjanian 1967] and [Browkin 1966] showed that no p-adic field $\mathbf{Q}_p$ is C_2. For some years it remained open whether p-adic fields $\mathbf{Q}_p$ and their finite algebraic extensions $K_\mathfrak{p}$ could be C_3. But by work of Arkhipov and Karatsuba (1981), Lewis and Montgomery (1983) and finally Alemu [1985] we now know that $K_\mathfrak{p}$ is not C_i for all $i \in \mathbf{N}$. So we could say that Artin's Conjecture is *very false*.

(c) On the other hand [Ax-Kochen 1965] proved by using methods from model theory that Artin's Conjecture is *almost true* in the following sense:
For fixed degree d there exists a finite (but non-constructible) set $S = S(d)$ of prime numbers such that any form f of degree d in $n > d^2$ variables over $\mathbf{Q}_p$ has a nontrivial zero provided $p \notin S$. It is even likely that $S(d) \subseteq \{p : p|d\}$.

Since we are particularly interested in quadratic forms and systems of quadratic forms in this book I should also mention the

(d) Restricted Artin Conjecture for systems of quadratic forms:
Let $(f_1(X), \ldots, f_r(X))$ be a system of r quadratic forms over $\mathbf{Q}_p$ where

$X = (X_1, \ldots, X_n)$, $n > 4r = r \cdot 2^2$. Is it true that $f_1 = \ldots = f_r = 0$ has a nontrivial zero?

This is true for $r = 1$ (classical), $r = 2$ [Demjanov 1956 and Birch–Lewis–Murphy 1962] and $r = 3$, $p > 49$ [Birch–Lewis 1965].

The general case is still open.

Of course similar questions can be asked if the underlying field K is a nonreal number field instead of a p-adic field. In more classical terms the condition K "nonreal" coincides with the condition K "totally imaginary", i.e. K has no imbedding into **R**. Recently Colliot-Thélène, Sansuc and Swinnerton-Dyer [1987] proved in two long and difficult papers:

2.6 Theorem. Let f_1, f_2 be two quadratic forms in at least nine variables over a nonreal number field K. Then the system $f_1 = f_2 = 0$ has a nontrivial solution in K.

You can also compare [Lang 1991] for more advanced methods, results and conjectures concerning this theorem and other C_i-problems.

As I said in Note 1.6(3) it is not known precisely when Artin's trick to go over from one form to a system of forms works. But we can easily show that it does not work if we restrict our consideration to forms of a *fixed* degree d, not even for $d = 2$ (in section 1 we had to allow all d or all d relatively prime to a given prime number p).

2.7 Example. Let K be the quadratic closure of **Q**. Then every quadratic form q in more than $1 = 2^0$ variables over K has a nontrivial zero in K since K is quadratically closed. But there exists a system (f_1, f_2) of two quadratic forms in $3 > 2 \cdot 2^0$ variables which has only the trivial zero, namely

$$f_1 = x^2 - yz, \quad f_2 = y^2 + xz + z^2.$$

PROOF. Assume that f_1, f_2 have a common nontrivial zero $(a, b, c) \in K^3$. $c = 0$ would give $a = 0$ and $b = 0$. Hence we may assume $c = 1$. Then $b = a^2$ and $b^2 + a + 1 = 0$, or $a^4 + a + 1 = 0$.

Using a little bit of Galois theory it is easy to show that the polynomial

$$p(x) = x^4 + x + 1 \in \mathbf{Q}[x]$$

is irreducible over **Q** and has Galois group $G = S_4$. Since $a \in K \subset \mathbf{C}$ is a zero of $p(x)$ it is clear that $\bar{a}$ (the complex conjugate of a) is another zero of $p(x)$. It is also clear that $\bar{a} \neq a$ since p has no real zero and that $\bar{a} \in K$. This implies

$$p(x) = (x - a)(x - \bar{a})(x^2 + a_1 x + a_0) \in K[x].$$

As K is quadratically closed there are no irreducible quadratic polynomials over K, hence

$$x^2 + a_1 x + a_0 = (x - a')(x - \bar{a}')$$

for some $a', \bar{a}' \in K$.

Thus we see that p splits over K, or in other words: K contains a splitting field L of p over $\mathbf{Q}$. But $[L : \mathbf{Q}] = |G| = 24$ whereas by definition K is an (infinite) composition of successive quadratic extensions which implies that every finite subextension of K has 2-power degree over $\mathbf{Q}$: contradiction.

This example leads to the following.

2.8 Definition. A field K is called a *quadratic C_0-field* if every system of r quadratic forms over K in $n > r$ variables has a nontrivial common zero in K (for every $r \in \mathbf{N}$).

The example shows that the condition "quadratically C_0" is stronger than the condition "quadratically closed". On the basis of this and other examples I put forward [Pfister 1979] the new suggestion:

2.9 Conjecture. Is every quadratic C_0-field K a p-field for some odd prime number p?

For char $K = 0$ this has been proved by [Leep 1990_1] using Galois theory and some very nice combinatorial arguments.

We conclude this section with a geometric application of Tsen's first theorem which has been pointed out to me by P.M.H. Wilson.

2.10 Example.

(1) Every smooth projective curve C of genus 0 over a C_1-field (or a quadratic C_1-field) L is rational.

(2) Theorem of Noether and Enriques: Let S be a smooth projective surface over an algebraically closed field K. Assume that S admits a surjective morphism $p : S \to C'$ onto a smooth curve C' such that the generic fibre $C = p^{-1}(x)$ is an irreducible curve of genus 0. Then S is a *ruled surface*, or in other words S is birationally equivalent to $C' \times \mathbf{P}^1$.

PROOF.

(1) Since C has genus 0 it is (up to isomorphism) a conic in projective 2-space $\mathbf{P}^2$ over L. Then C is the zero-set of a quadratic form f in three variables. As L is a C_1-field f has a nontrivial zero over L. This implies C is rational, the function field $L(C)$ of C is purely transcendental over L, say $L(C) = L(t)$.

(2) Put $L = K(C')$. This is a function field of transcendence degree 1 over K, hence a C_1-field by Tsen's theorem. Apply part (1) to the curve C over the field L. Then the function field $K(S)$ of S satisfies

$$K(S) = L(C) = L(t) = K(C')(t).$$

This shows that S is birationally equivalent to $C' \times \mathbf{P}^1$ over K.

For more details see Chapter 1 of the book of Šafarevič: Algebraic Surfaces [1965].

Chapter 6

Hilbert's 17th Problem

§1. Preliminaries on Ordered and Real Fields

At the beginning of Chapter 3 a field K was defined to be (formally) real if -1 is not a sum of squares in K, or, equivalently, if a nontrivial sum of squares is never zero. This suggests we study the subset of all *sums of squares* of a field K or a commutative ring A in more detail.

1.1 Notation. Let A be a commutative ring (with identity element $1 \neq 0$). We put

$$\sum A = \{a \in A : a \text{ is a sum of squares in } A\},$$
$$\sum A^\bullet = \sum A \setminus \{0\}.$$

The following result is trivial:
1.2 Lemma.

(1) $\sum A$ is closed under addition and multiplication.

(2) For a field K the set $\sum K^\bullet$ is a multiplicative group.

(3) K formally real $\iff -1 \notin \sum K$.

The main definition of this section is as follows:
1.3 Definition.

(1) A subset P of a commutative ring A is called a *preorder(ing)* of A if

$$P + P \subseteq P, \quad P \cdot P \subseteq P, \quad \sum A \subseteq P, \quad -1 \notin P.$$

(2) A preorder(ing) P is called an *order(ing)* of A if

$$P \cup -P = A, \quad P \cap -P = \{0\}.$$

From this definition we see immediately that either $-1 \in \sum A$ (i.e. A has finite level by Definition 2.1 of Chapter 3) or $\sum A$ is the *smallest preorder* of A.

1.4 Lemma. Let P be a preorder of A and let $a, b \in A$. If $ab \in P$ then $P + aP$ or $P - bP$ is a preorder of A.

PROOF. It is clear that $P + aP$ and $P - bP$ are closed under additon and multiplication. Both sets contain P, hence $\sum A$. We have to show $-1 \notin P + aP$ or $-1 \notin P - bP$. Assume $-1 = p_1 + ap_2 = p_3 - bp_4$ where $p_1, \ldots, p_4 \in P$. Then $(ap_2)(-bp_4) = (1 + p_1)(1 + p_3) = 1 + p_1 + p_3 + p_1 p_3$, $-1 = p_1 + p_3 + p_1 p_3 + (ab)p_2 p_4 \in P$: contradiction.

1.5 Lemma. Let P be a *maximal* preorder of A (i.e. maximal with respect to inclusion). Then $P \cup -P = A$ and $P \cap -P = \mathfrak{p}$ is a prime ideal of A. If $A = K$ is a field, then $\mathfrak{p} = 0$ and P is an order of K.

PROOF. Let $a \in A$ be arbitrary. Put $b = a$ in the last lemma. Then $ab = a^2 \in P$, hence $P + aP$ or $P - aP$ is a preorder. By the maximality of P this implies $a \in P$ or $-a \in P$. Hence $P \cup -P = A$. $P \cap -P$ is closed under addition. Suppose $a \in P \cap -P$, $b \in A = P \cup -P$. Then $ab \in P \cap -P$. Hence $\mathfrak{p} = P \cap -P$ is an ideal of A. Finally let $a, b \in A$ such that $ab \in \mathfrak{p}$. If $a \notin \mathfrak{p}$, say $a \notin P$, then $P + aP$ is strictly bigger than P hence cannot be a preorder (by the maximality of P). Thus Lemma 1.4 implies that $P - bP$ is a preorder, hence $-b \in P$ by the maximality of P. The same reasoning applies for the element $-b$ instead of b since also $a(-b) = -ab \in \mathfrak{p}$. This shows $b \in P \cap -P = \mathfrak{p}$, i.e. $\mathfrak{p}$ is a prime ideal. The last claim for $A = K$ is clear.

Whenever $-1 \notin \sum A$, $P_0 = \sum A$ is a preorder of A. In this case the existence of at least one maximal preorder is guaranteed by Zorn's lemma.

In particular we get the following.

1.6 Corollary. (Artin–Schreier). Let K be a field, then

$$K \text{ formally real} \iff K \text{ has an order } P.$$

In a real field the relation between preorders and orders is given by

1.7 Proposition. Let T be a preorder of K. Then $T = \cap P$ where the intersection is taken over all orders P containing T.

PROOF. $T \subseteq \cap P$ is clear. Let $a \in K \backslash T$. Then $T - aT$ is a preorder since $-1 = t_1 - at_2$ with $t_1, t_2 \in T$ leads to the contradiction $a = \frac{t_1}{t_2} + \frac{1}{t_2} = (\frac{1}{t_2})^2(t_1 t_2 + t_2) \in T$. By 1.5 there is an order P containing $T - aT$. This shows $-a \in P$, $a \notin P$ (since $P \cap -P = 0$ and $a \neq 0$). Therefore $\cap P = T$.

1.8 Corollary. The intersection $\cap P$ of *all* orders P of K is exactly the set $P_0 = \sum K$.

1.9 Note.

(1) An ordering P of K induces a binary relation $\leq$ on K defined by

$$a \leq b :\Longleftrightarrow b - a \in P.$$

This relation satisfies the following rules:

(i) $a \leq a$,

(ii) $a \leq b$ and $b \leq a \Longrightarrow a = b$,

(iii) $a \leq b$ or $b \leq a$,

(iv) $a \leq b$ and $b \leq c \Longrightarrow a \leq c$,

(v) $a \leq b \Longrightarrow a + c \leq b + c$,

(vi) $a \leq b$, $0 \leq c \Longrightarrow ac \leq bc$.

Here (i)–(iv) imply that K is totally ordered by $\leq$, (v) says that the order is linear w.r.t. $+$, (vi) says that the order respects multiplication by non-negative elements. (i)–(vi) are the classical axioms for an *order relation* $\leq$ on a field K.

(2) Conversely every relation $\leq$ on K which satisfies (i)–(vi) leads to the subset $P := \{a \in K : 0 \leq a\}$ which satisfies the conditions (1) and (2) of Definition 1.3. P is sometimes called the *positive cone* of $\leq$.

(3) This allows us to switch freely from P to $\leq$ and to call a pair (K, P) or $(K, \leq)$ an *ordered field*.

(4) Finally it is convenient to write

$$a < b \iff a \leq b \quad \text{and} \quad a \neq b,$$
$$b > a \iff a < b.$$

(5) Corollary 1.8 says that an element $a \in K^{\bullet}$ is a sum of squares if and only if a is *totally positive*, i.e. $a \geq 0$ for every order relation $\leq$ on K.

1.10 Examples.

(1) **Q** and **R** have only one order, namely the usual order.

(2) The rational function field $K = \mathbf{R}(t)$ has the following orders:

a) For each $r \in \mathbf{R}$ the order $P_{r,+}$ where $t > r$ and $t < s$ for every $s \in \mathbf{R}$, $s > r$.

b) For each $r \in \mathbf{R}$ the order $P_{r,-}$ where $t < r$ and $t > s$ for every $s \in \mathbf{R}$, $s < r$.

c) The orders $P_{\infty,+}$ where $t > s$ for every $s \in \mathbf{R}$ and $P_{\infty,-}$ where $t < s$ for every $s \in \mathbf{R}$.

PROOF.

(1) For $k = \mathbf{R}$ we have $\mathbf{R}_+ = \{r \in \mathbf{R} : r \geq 0 \text{ in the usual order}\} = \mathbf{R}^2 = \sum \mathbf{R}$ which shows that $P_0 = \sum \mathbf{R}$ is an order. An arbitrary order P satisfies $P_0 \subseteq P$. This implies $P_0 = P$ by 1.3 (2). Nearly the same proof works for $\mathbf{Q}$: Every natural number n is a sum of squares, since $n = \underbrace{1^2 + \ldots + 1^2}_{n}$.

(Actually n is a sum of four squares by the famous theorem of Lagrange.) This implies that every positive (in the usual order) rational number $q = \frac{n}{m} = \frac{n \cdot m}{m^2}$ ($m, n \in \mathbf{N}$) is a sum of squares. Therefore $P_0 = \sum \mathbf{Q}$ is again an order of $\mathbf{Q}$, hence the only one.

(2) Let P be an order of $K = \mathbf{R}(t)$. Look at $\sup\{r' \in \mathbf{R} : r' < t$ w.r.t. P, i.e. $t - r' \in P\}$. This supremum can be $-\infty$ (if the set under consideration is empty) or a real number r or $+\infty$ (if $r' < t$ for all $r' \in \mathbf{R}$). The first, resp. third, case lead to the orders $P_{\infty,-}$, resp. $P_{\infty,+}$, the second case to the order $P_{r,-}$ if $t < r$ or $P_{r,+}$ if $t > r$. (Note that $t \neq r$ since $t \notin \mathbf{R}$.) It remains to show that the above weak description of the sets $P_{r,-}$, $P_{r,+}$, $P_{\infty,-}$, $P_{\infty,+}$, which gives only the place of the single element t with respect to the line $\mathbf{R}$ in the linear order of K, suffices for a full description of these sets. Since $\mathbf{R}(t) = \mathbf{R}(t - r) = \mathbf{R}(-t) = \mathbf{R}(t^{-1})$ it is enough to do this for the case $P_{0,+}$ in which $t > 0$ but $t < s$ for every $s \in \mathbf{R}$, $s > 0$, i.e. t is *infinitely small*. Consider $f(t) = a_k t^k + a_{k+1} t^{k+1} + \ldots + a_n t^n \in \mathbf{R}[t]$ such that $a_k \neq 0$. Then

$$f(t) \in P_{0,+} \cap \mathbf{R}[t] :\Longleftrightarrow a_k > 0.$$

The full set $P_{0,+}$ is then given by

$$P_{0,+} = \{0\} \cup \left\{ \frac{f(t)}{g(t)} \in K : 0 \neq f(t) \in \mathbf{R}[t], \right.$$

$$\left. 0 \neq g(t) \in \mathbf{R}[t], \ f(t)g(t) \in P_{0,+} \cap \mathbf{R}[t] \right\}.$$

It is nearly trivial to check that $P_{0,+}$ is actually an order of K, and a continuity argument for polynomials $f(t) = a_k t^k + \ldots + a_n t^n$ ($k \leq n$) as above shows that $P_{0,+}$ is the only order of K for which t is positive and infinitely small.

Let (K, P) and (K', P') be ordered fields with $K \subset K'$. Then (K', P') is called an *extension* of (K, P) if $P' \cap K = P$. For algebraic extensions the following result is fundamental:

1.11 Theorem. Let (K, P) be an ordered field.

(1) If $K' = K(\sqrt{d})$ with $d \in K$ there is an extension P' of P to K' if and only if $d \in P$.

(2) If $[K' : K] = n$ is odd there is always an extension of P to K'.

PROOF.

(1) Since $d = (\sqrt{d})^2 \in P'$ for every order P' of K' the condition on d is necessary. Assume now $d \in P$, but $d \notin K^2$. Consider the set $T' \subset K'$ consisting of all finite sums

$$\sum c_i \alpha_i^2$$

with $c_i \in P$, $\alpha_i \in K'$. T' is closed under addition and multiplication and contains all squares of K'. T' is a preorder since otherwise we could find an equation

$$-1 = \sum c_i \alpha_i^2 = \sum c_i (a_i + b_i \sqrt{d})^2 \quad (a_i, b_i \in K)$$

which leads to the contradiction

$$-1 = \sum c_i (a_i^2 + b_i^2 d) \in P.$$

By Zorn's lemma there exists a maximal preorder P' of K' containing T', and by 1.5 P' is an order of K'. Since $P' \supseteq T' \supseteq P$ it follows that $P' \cap K = P$.

(2) As in (1) we consider the set $T' \subset K'$ consisting of all finite sums $\sum_1^m c_i \alpha_i^2$ with $c_i \in P$, $\alpha_i \in K'$, $m \in \mathbb{N}_0$. It suffices to show $-1 \notin T'$, or in other words: The quadratic form $\langle c_1, \ldots, c_m \rangle$ does not represent -1 over K'. Clearly we may assume $c_1, \ldots, c_m \in P \backslash \{0\}$. Then $\varphi = \langle 1, c_1, \ldots, c_m \rangle$ is obviously anisotropic over K and we have to show that φ remains anisotropic over K'. This follows from the following well-known result

1.12 Theorem. (T.A. Springer) Let L/K be a field extension of odd degree. Let φ be an anisotropic quadratic or symmetric bilinear form over K. Then the induced form φ_L over L is anisotropic, too. In particular, the natural map $W(K) \to W(L)$ is injective.

PROOF. We use induction on the degree $n = [L : K]$, the case $n = 1$ being trivial. By repeated application of the theorem (if necessary) we can also assume that $L = K(\alpha)$ is a simple extension. Let $p(X) \in K[X]$ be the minimal polynomial of α. Every element $\alpha_i \in L$ has the form

$$\alpha_i = g_i(\alpha), \quad \text{where} \quad g_i(X) \in K[X], \deg g_i < n = \deg p.$$

Let now $\varphi(x_1,\ldots,x_m) = \sum_{i,j=1}^{m} a_{ij}x_i x_j$ be an anisotropic form over K and assume $\varphi(\alpha_1,\ldots,\alpha_m) = 0$ over L where not all α_i vanish. Then the polynomial

$$0 \neq \varphi(g_1(X),\ldots,g_m(X)) \in K[X]$$

has even degree at most $2(n-1)$ and is divisible by $p(X)$. We get

$$\varphi(g_1(X),\ldots,g_m(X)) = p(X) \cdot q(X)$$

where $q(X) \in K[X]$ has odd degree $\leq n - 2$. In addition we can assume that $g_1,\ldots,g_m$ have no common (non-constant) factor f since otherwise f^2 could be cancelled in the above equation. The polynomial $q(X)$ has at least one irreducible factor $r(X)$ of odd degree. Let $r(\beta) = 0$, $L' = K(\beta)$. Then $[L' : K]$ is odd and $< n$, $(g_1(\beta),\ldots,g_m(\beta)) \neq (0,\ldots,0)$ and $\varphi(g_1(\beta),\ldots,g_m(\beta)) = 0$, i.e. φ is isotropic over L'. This contradicts the induction hypothesis.

The basis for Hilbert's 17th problem is the real closed fields which we will introduce now.

1.13 Definition. A formally real field R is called *real-closed* if R has no proper algebraic real extension field.

In other words: There is no field L between R and an algebraic closure $\bar{R}$ of R which is real and different from R.
From 1.11 we immediately deduce:

1.14 Theorem. Let R be real-closed.

(1) Every polynomial $f(X) \in R[X]$ of odd degree has a zero in R.

(2) $P_0 = R^2$ is the only order of R.

(3) $C = R(i)$ with $i^2 = -1$ is algebraically closed.

PROOF.

(1) We may suppose that f is irreducible. If $\deg f > 1$ and α is a zero of f in $\bar{R}$ then $L = R(\alpha)$ is an extension field of odd degree > 1. Since R is real it has at least one order P. By 1.11 this order can be extended to an order P' of L which shows in particular that L is real: contradiction.

(2) Let P be an order of R, let $a \in P$, Then $R(\sqrt{a})$ can be ordered by 1.11. Thus we must have $R(\sqrt{a}) = R$, i.e. $a \in R^2$. This implies $P = R^2 = P_0$ and the uniqueness of the order of R.

(3) Since $R = P \cup -P$ and $P \cap -P = 0$ with $P = R^2$ the field R has exactly one quadratic extension, namely $C = R(i)$. We have to show that $C = \bar{R}$ is algebraically closed. Let E be a finite extension of C. By further extending E (if necessary) we may suppose that E/R is a Galois extension with Galois group G. (Note that char $R = 0$.) Let H be a Sylow 2-subgroup of G, let R' be the fixed field of H. Then $[R' : R] = [G : H]$ is odd. Hence we must have $R' = R$, $G = H$. This implies that G is a 2-group. Consider now the group $F = \mathrm{Gal}(E/C)$ which is also a 2-group. Assume $E \neq C$, i.e. $F \neq \{1\}$. Then F has a subgroup F' of index 2 and the fixed field E' of F' is a quadratic extension of C, say $E' = C(w)$ with $w^2 = z \in C$. We will see that this is impossible because C is quadratically closed:

Let $z = a + bi$ with $a, b \in R$. By part (2) there exists an element $c \in P = R^2$ such that $c^2 = a^2 + b^2$. Since $c + a \geq 0$ and $c - a \geq 0$ there are elements $d_1, d_2 \in P$ with

$$d_1^2 = \frac{c+a}{2}, \quad d_2^2 = \frac{c-a}{2}.$$

Put $w = d_1 + \varepsilon d_2 i$ where $\varepsilon = 1$ for $b \geq 0$, $\varepsilon = -1$ for $b < 0$. Then

$$w^2 = d_1^2 - d_2^2 + 2\varepsilon d_1 d_2 i = a + \varepsilon \sqrt{c^2 - a^2} i = a + \varepsilon |b| i = a + bi.$$

The proof is finished.

By practically the same arguments one can prove the partial converse of 1.14: Let (K, P) be an ordered field with the following two properties:

- Every $f(X) \in K[X]$ of odd degree has a zero in K.

- Every element $a \in P$ is a square in K.

Then K is real-closed and $P = K^2$.

Since the field **R** of real numbers satisfies these two properties 1.14 includes the classical result: $C = \mathbf{R}(i)$ is algebraically closed.

Our last result in this section concerns the relation between ordered fields and real-closed fields.

1.15 Definition. Let (K, P) be an ordered field. A real closed field R is called the *real-closure* (or real-closed hull) of (K, P) if R/K is algebraic and $R^2 \cap K = P$, i.e. the unique order R^2 of R extends the given order P of K.

1.16 Theorem. Every ordered field (K, P), and thus in particular every real field K, has a real closure R.

PROOF. Let $\bar{K}$ be a fixed algebraic closure of K. By repeated application of 1.11(1) we adjoin the square roots $\sqrt{a} \in \bar{K}$ of all elements $a \in P$ to K. This yields an ordered extension field (K', P') of (K, P) such that $P = K'^2 \cap K$. (In general the extension K'/K will be infinite but of course algebraic since $K' \subseteq \bar{K}$.) Since K' is (formally) real we can apply Zorn's lemma to the set of all *real* fields between K' and $\bar{K}$ (ordered by inclusion). Let R be a *maximal* element of this set. Definition 1.13 shows that R is real-closed. Finally $R^2 \cap K \supseteq K'^2 \cap K = P$ and $(-R^2) \cap K \supseteq (-K'^2) \cap K = -P$, hence $R^2 \cap K = P$.

1.17 Note. It can be shown that two real closures $(R_1, R_1^2), (R_2, R_2^2)$ of (K, P) are necessarily isomorphic as ordered fields. Hence we can speak of *the real-closure* of an ordered field. The proof of this fact is considerably more involved than the previous proofs and is omitted.

1.18 Example. The real closure of $\mathbf{Q}$ (with its unique order) is the field $\mathbf{R}_a$ of all *real algebraic* numbers, $\mathbf{R}_a \subset \mathbf{R}$.

1.19 Note. Though we have used Zorn's lemma several times it is possible to prove the existence and uniqueness of a real closure of an arbitrary ordered field without it (of course at the expense of a much longer proof). This is essentially done by reduction to the case of countable fields. In contrast to this it is not possible to prove the existence of the algebraic closure $\bar{K}$ of an arbitrary field K without Zorn's lemma or equivalent tools from set theory like the axiom of choice or the well-ordering principle. For details see [Sander 1991].

Historical Note. The abstract theory of ordered fields, formally real fields and real-closed fields was developed in the fundamental paper [Artin–Schreier 1927]. It forms part of most algebra books, e.g. [Jacobson 1980] or [Lorenz 1990]. A full account is also given in [S, Chapter 3]. Our presentation closely follows the last two books.

§2. Artin's Solution and other Qualitative Results

In his famous address at the 1900 International Congress of Mathematicians in Paris David Hilbert proposed a list of 23 unsolved problems which he considered to be important for the future development of mathematics. No. 17 of these problems (in a slightly modernized version) reads as follows:

2.1 Hilbert's Problem. Let $K = \mathbf{R}(X_1, \ldots, X_n)$. Call an element $f = f(X_1, \ldots X_n) \in K$ *positive semi-definite* if $f(a) \geq 0$ for all $a = (a_1, \ldots, a_n) \in \mathbf{R}^n$ where f is defined. Is it true that f is a sum of squares in K?

Hilbert himself studied this and related questions since 1888. In [Hilbert 1888] he showed that the corresponding question for the polynomial ring

$\mathbf{R}[X_1, \ldots, X_n]$ has a negative answer for $n \geq 2$. We have already seen an explicit example in 2.4 of Chapter 1. In [Hilbert 1893$_1$] he answered the above problem for the case $n = 2$ using deep methods about abelian functions. In fact he showed that four squares suffice for every positive semi-definite $f \in \mathbf{R}(X_1, X_2)$.

Of course the same problem can be asked if the constant field $\mathbf{R}$ is replaced by another real-closed field R or even an arbitrary ordered field (k, P) which has a unique order. This last condition is necessary since the constant positive functions $f = a > 0$, $a \in k$, must be sums of squares, i.e. we must have $P = \sum k = P_0$. This implies that the order P of k is unique. A prominent example is the field $k = \mathbf{Q}$.

Before we outline Artin's proof of Hilbert's problem we consider the converse question whether a rational function $f \in K$ which can be represented as a sum of squares must be positive semi-definite. This is nearly trivial but not entirely.

2.2 Lemma. Let (k, P) be an ordered field, let $K = k(X_1, \ldots, X_n)$ and let $f \in \sum K$ be a sum of squares. Then $f(a) \geq 0$ for every $a = (a_1, \ldots, a_n) \in k^n$ at which f is defined.

PROOF. Let $f = \sum(\frac{g_j}{h})^2$ with $g_j, h \in k[X_1, \ldots, X_n]$, $h \neq 0$. By the linear transformation $X_i \to X_i + a_i$ $(i = 1, \ldots, n)$ we may assume $a = (0, \ldots, 0)$. f is defined at $a = 0$ if there is (at least one) representation $f = \frac{r}{s}$ with $r, s \in k[X_1, \ldots, X_n]$ and $s(0) \neq 0$. We can assume $s(0) > 0$. Then we have to show $r(0) \geq 0$. If $h(0) \neq 0$ then $r(0) = s(0) \cdot \sum(\frac{g_j(0)}{h(0)})^2 \geq 0$. This is the trivial case. If $h(0) = 0$ we must be a little more careful: Consider the identity

$$r(X) \cdot h(X)^2 = s(X) \cdot \sum g_j(X)^2$$

and assume $r(0) < 0$. Since $h \neq 0$ but $h(0) = 0$ we have

$$h = h_d + h_{d+1} + \cdots$$

where h_δ is homogeneous of degree δ and $h_d \neq 0$, $d > 0$. Write $X_i = TY_i$. Then $h_\delta(X_1, \ldots, X_n) = T^\delta h_\delta(Y_1, \ldots, Y_n)$. Since $h_d \neq 0$ and since k is infinite (because char $k = 0$) there exists a point $b = (b_1, \ldots, b_n) \in k^n$ such that $h_d(b) \neq 0$. Substitute $X_i = b_i T$ in the above identity. We get

$$(r(0) + T \cdot *)(T^d \cdot h_d(b) + T^{d+1} \cdot *)^2 = (s(0) + T \cdot *) \sum g_j(Tb)^2.$$

Compare the coefficients of the lowest non-vanishing term in this identity: We must have $T^d | g_j(Tb)$ for all j. Furthermore the coefficient of T^{2d} on the left is $r(0)h_d(b)^2 < 0$ whereas the coefficient on the right is $s(0)$ times a sum of squares, hence ≥ 0: contradiction.

The result which E. Artin proved in 1927 and which in particular solves Hilbert's problem is the following.

2.3 Theorem. Let (K_0, P_0) be an ordered field with real closure R_0. Let $K = K_0(X_1, \ldots, X_n)$ be the rational function field in n variables over K_0 and let $f \in K$ be a rational function such that
(1_s) $f(a) = f(a_1, \ldots, a_n) \geq 0$ in R_0 for all $a = (a_1, \ldots, a_n) \in R_0^n$ where $f(a)$ is defined.
Then there exist finitely many $p_i \in P_0 \subset K_0$ and $f_i \in K$ such that
(2) $f = \sum p_i f_i^2$.
If in particular $P_0 = \sum K_0$ is the only order of K_0 then the p_i are sums of squares (in K_0) and f is a sum of squares (in K).

Artin's own proof used specialization arguments and Sturm's theorem on counting the real zeros of a polynomial over **R** in a given interval, and of course the Artin–Schreier theory of section 1 which obviously was a "by-product" of the attempt to solve Hilbert's problem. Later, S. Lang [1953] replaced the specialization methods by more systematic investigations on real valuations and real places. He showed that Theorem 2.3 can be proved relatively easily from the following technical result:

2.4 Artin–Lang Homomorphism Theorem. Let R be a real-closed field, let $A = R[x_1, \ldots, x_n]$ be a finitely generated R-algebra which has no zero-divisors (the x_i need not be algebraically independent) such that the quotient field $K = R(x_1, \ldots, x_n)$ of A is formally real. Then there exists an R-algebra homomorphism

$$\varphi : R[x_1, \ldots, x_n] \to R.$$

Though the proof of this theorem has been further simplified by work of Prestel and Lorenz it is still rather complicated and lengthy. Therefore I refer the interested reader to the presentation in [S, Ch. 3 Theorem 3.1] or [Lorenz 1990, §20, Satz 10].

PROOF of 2.3 from 2.4:

(1) Consider the set T consisting of all finite sums of the shape $\sum p_i f_i^2$ with $p_i \in P_0$, $f_i \in K$. Obviously T is a preorder in K. The fact $-1 \notin T$ follows from 2.2. By 1.7 we have $T = \cap P$ where P runs through all orders of K which contain T. Therefore our claim $f \in T$ follows if we can show $f \in P$ for every $P \supseteq T$.

(2) Assume that there exists an order P containing T such that $f \notin P$, i.e. $f \neq 0$ and $-f \in P$. Let R be a real closure of (K, P) which exists by 1.16. Then $-f$ is a square in R, say $-f = w^2$, $0 \neq w \in R$. The considerations after the proof of 1.14 show that the set $R_0' = \{\alpha \in R : \alpha \text{ algebraic over } K_0\}$ is a real closure of (K_0, P_0). Since the real closure R_0 of (K_0, P_0) is essentially unique by 1.17 the condition (1_s) in 2.3 does not depend on the choice of R_0. This allows us to assume $R_0 = R_0' \subseteq R$.

(3) Let $f = \frac{g}{h}$ with $g, h \in K_0[X_1, \ldots, X_n]$ be a fixed representation of f as a quotient of two polynomials. We apply 2.4 to the R_0-algebra

$$A = R_0[X_1, \ldots, X_n, \frac{1}{h}, w, \frac{1}{w}] \subseteq R_0(X_1, \ldots, X_n, w) \subseteq R$$

and fix a homomorphism $\varphi : A \to R_0$. Put $a_i = \varphi(X_i)$, $i = 1, \ldots, n$. Then

$$\varphi(f) = \frac{\varphi(g)}{\varphi(h)} = \frac{g(a_1, \ldots, a_n)}{h(a_1, \ldots, a_n)} = f(a_1, \ldots, a_n)$$

is defined since the existence of $\varphi(\frac{1}{h})$ implies $\varphi(h) \neq 0$. Similarly the equation $\varphi(w)\varphi(\frac{1}{w}) = \varphi(1) = 1$ implies $\varphi(w) \neq 0$. Finally the equation $f = -w^2$ implies

$$f(a_1, \ldots, a_n) = \varphi(f) = -\varphi(w)^2 < 0 \quad \text{in} \quad R_0.$$

This is a contradiction to (1_s) and proves the theorem.

Note. Since (1_s) or (2) holds for a rational function $f = \frac{g}{h}$ if and only if it holds for the polynomial gh it would be enough to prove 2.3 for polynomials.

If K_0 is not real-closed then the strong condition (1_s) that f be positive semi-definite on R_0^n seems to be slightly artificial since all other properties in Theorem 2.3 depend only on (K_0, P_0) and n. Thus it is natural to ask whether (1_s) could be replaced by the weaker condition
(1_w) $f(a) = f(a_1, \ldots, a_n) \geq 0$ in K_0 for all $a = (a_1, \ldots, a_n) \in K_0^n$ where $f(a)$ is defined
(i.e. f is positive semi-definite on K_0^n).
Concerning this question we have

2.5 Theorem. [McKenna 1975] (1_w) implies (2) if and only if K_0 is *dense* in R_0 with respect to the order topology on R_0. (This is always the case if the order P_0 on K_0 is archimedean (Compare the Notes at the end of Chapter 4).)

One direction of this theorem is pretty clear. Namely, if K_0 is dense in R_0, then the continuity of a rational function f in a point a where it is defined implies that (1_w) for f induces (1_s) for f. The other direction is less obvious. A further generalization of McKenna's theorem has been considered in two papers by Z. Guangxin [1988, 1991].
We shall not give the proof of Theorem 2.5 here but rather include an example, which shows that already for $n = 1$ there may exist rational functions which satisfy (1_w) but take *negative* values for suitable points $b \in R_0 \backslash K_0$ if K_0 is not dense in R_0. Then of course condition (2) in 2.3 cannot hold.
The first example of this kind was given in [Dubois 1967]. An easier example can be found in the book of [Lorenz 1990]. I repeat it here.

2.6 Example. Let $K_0 = \mathbf{Q}(t)$ where t is transcendental, let P_0 be the order of K_0 in which $t > 0$ and infinitely small. Then $R_0 = \bigcup_{n \in \mathbf{N}} \mathbf{R}_a((t^{1/n}))$ is the field of real algebraic Puiseux series over $\mathbf{R}_a$. (Compare Note (2b) at the end of Chapter 4.) Let $K = K_0(X)$, $f(X) = X^4 - 5tX^2 + 4t^2$. Then f is positive semi-definite on K_0 but $f(a) < 0$ for $a = \sqrt{\frac{5}{2}t} \in R_0$.

PROOF. Firstly we show that there is no $a \in K_0$ with $t < a^2 < 4t$ (for P_0). Assume that we have such an a. Then w.l.o.g. $a > 0$. Look at the Laurent series expansion of a:

$$
\begin{aligned}
a &= qt^n + q't^{n+1} + \dots \quad (n \in \mathbf{Z};\ q, q', \dots \in \mathbf{Q},\ q > 0),\\
-t + a^2 &= -t + q^2 t^{2n} + \dots > 0 \quad \Longrightarrow n \le 0,\\
4t - a^2 &= 4t - q^2 t^{2n} - \dots > 0 \quad \Longrightarrow n > 0.
\end{aligned}
$$

This contradiction proves that there is no a^2 in the open interval $(t, 4t)$ and consequently no $a \in K_0$ in the open interval $(\sqrt{t}, 2\sqrt{t})$ of R_0.

Assume now that there is an $a \in K_0$ with $f(a) < 0$. Since $f(X) = (X^2 - \frac{5}{2}t)^2 - (\frac{3}{2}t)^2$ this would imply $-\frac{3}{2}t < a^2 - \frac{5}{2}t < \frac{3}{2}t$, i.e. $t < a^2 < 4t$: contradiction.

Another generalization of Artin's theorem 2.3 is towards algebraic function fields instead of rational function fields. At the same time it reveals the more geometric aspects of the theory. For simplicity we suppose that K_0 has a unique order $P_0 = \sum K_0$. Let R_0 be the real closure of K_0 as before. Consider a non-empty irreducible affine real algebraic variety $V \subset R_0^n$ with vanishing (prime) ideal $I_{R_0}(V) = \{f \in R_0[X_1, \dots, X_n] : f(V) = 0\}$. Suppose that V is defined over K_0, i.e. $I_{R_0}(V)$ is generated by (finitely many) $f_i \in K_0[X_1, \dots, X_n]$. Then $I(V) = I_{R_0}(V) \cap K_0[X_1, \dots, X_n]$ is also prime. The integral domain $A = K_0[V] = K_0[X_1, \dots, X_n]/I(V)$ is called the *affine algebra* of V over K_0. Its quotient field $K = \mathrm{quot}(A)$ is called the *function field* of V over K_0. Elements of A may be considered as R_0-valued functions on V (by substituting a point $r = (r_1, \dots, r_n) \in R_0^n$ for the n-tuple $(X_1, \dots, X_n)$ of variables). Similarly an element $f = \frac{g}{h} \in K$ (with $g, h \in A$) may be considered as a function on V which is defined at points r where $h(r) \ne 0$. $f \in K$ is called *positive semi-definite on* V if $f(r) \ge 0$ for all $r \in V$ where f is defined. (For the affine space $V = R_0^n$ we have $I(V) = 0$ and $K = K_0(X_1, \dots, X_n)$ which is the field considered in 2.3.)

With this terminology we have the following.

2.7 Theorem. The following statements for an element $f \in A$ are equivalent:

(1) $f(r) \ge 0$ for all *regular* points $r \in V$.

(2) $f(r) \ge 0$ for all $r \in U \subseteq V$ where U is a non-empty *Zariski-open* subset of V.

(3) $f \in \sum K$, i.e. f is a sum of squares in K.

For a proof see [Lorenz 1990, §21, Satz 2] or [Bochnak–Coste–Roy 1987, Theorem 6.1.9]. The terms "regular" and "Zariski-open" are defined as usual in algebraic geometry.

You may be surprised that property (3) in Theorem 2.7 follows from property (1) or (2) without knowing whether $f(r) \geq 0$ for *all* $r \in V$. This seems to allow the possibility that $f \in \sum K$ takes negative values at certain (necessarily *singular*) points of V. We shall see that this phenomenon can actually occur (Lemma 2.2 tells us that it does not happen for the full space $V = R_0^n$ which has no singular points).

2.8 Example. Let $K_0 = R_0 = \mathbf{R}$, let $V \subset \mathbf{R}^2$ be the zero set of the polynomial

$$f(X, Y) = Y^2 - (X^3 - X^2) \in \mathbf{R}[X, Y].$$

Then V is a real variety of dimension 1. The function $x - 1 \in A = \mathbf{R}[V]$ takes the value -1 at the (singular) point $(0, 0) \in V$. Nevertheless $x - 1 = (\frac{y}{x})^2$ is a square in $K = \mathrm{quot}(A)$.

PROOF. Obviously f is an irreducible element of the factorial domain $\mathbf{R}[X, Y]$. Hence the principal ideal (f) is a prime ideal. Every point $(x, y) \in V \backslash (0, 0)$ satisfies $x \neq 0$, $x^3 = y^2 + x^2$; $x = (\frac{y}{x})^2 + 1$. Put $t = \frac{y}{x} \in \mathbf{R}$. Then $x = t^2 + 1$, $y = tx = t(t^2 + 1)$. Conversely, every point

$$(*) \qquad\qquad x = t^2 + 1, \quad y = t(t^2 + 1), \quad t \in \mathbf{R},$$

belongs to V.

Consider now a polynomial $g = g(X, Y) \in \mathbf{R}[X, Y]$ which vanishes on V. By the division algorithm in $\mathbf{R}(X)[Y]$ we get

$$g(X, Y) = h(X, Y)f(X, Y) + u(X)Y + v(X)$$

where $h(X, Y) \in \mathbf{R}[X, Y]$ and $u(X)$, $v(X) \in \mathbf{R}[X]$. Substituting $(*)$ into this equation we see that

$$0 = 0 + t(t^2 + 1)u(t^2 + 1) + v(t^2 + 1)$$

for all $t \in \mathbf{R}$. Comparing degrees shows $u = v = 0$. This proves $g \in (f)$ and $I(V) = (f)$. Hence V is an irreducible variety of dimension 1. Actually $(*)$ shows in addition that $K = \mathbf{R}(V) = \mathbf{R}(T)$ is a rational function field, i.e. V is a *rational curve*.

Putting $x = X \bmod f \in A$, $y = Y \bmod f \in A$ the last two statements are clear.

Let us now mention some other results which are intimately related to Artin–Schreier theory and Hilbert's 17th problem. Real-closed fields are of great interest in mathematical logic, or more precisely first order predicate calculus and model theory. Let R_0 be real-closed, let $f \in R_0[X_1, \ldots, X_n]$. Then statements like

(1) $$(\forall x_1 \ldots \forall x_n) f(x_1, \ldots, x_n) \geq 0$$

and

(2) $$(\exists g_0 \ldots \exists g_r) f g_0^2 = g_1^2 + \ldots + g_r^2, \quad g_0 \neq 0$$

with $g_0, \ldots, g_r \in R_0[X_1, \ldots, X_n]$, $\deg g_j \leq m$ $(j = 0, \ldots, r)$, are first order sentences for every pair of natural numbers r and m. (First order means that the quantifiers $\forall, \exists$ should only be taken over elements in R_0, not over subsets of R_0, therefore we have to fix m and r.)

Now the famous "Tarski principle" tells us that the first order theory of real-closed fields allows "quantifier elimination" and is "model complete". In particular, a positive semi-definite polynomial f satisfies (1) in every real closed field R containing R_0, e.g. any real closure R of the field $R_0(X_1, \ldots, X_n) = K$. Hence $f \in P$ for every order P of K from which $f \in \sum K$, i.e. statement (2) for suitable m and r follows immediately. This indicates the main idea of A. Robinson's model theoretic solution of Hilbert's 17th problem, [Robinson 1955]. Later Kreisel [1957] gave a constructive proof and showed that there is an upper bound for the number r of squares, depending only on n and the degree of f (but not on the coefficients of f). For the interplay between mathematical and logical methods and for an impressive list of new results I refer the interested reader to the marvellous book of Bochnak, Coste and Roy.

Let me close this section by mentioning an old elegant paper of Habicht [1940] who proves that for a *strictly definite* homogeneous $f \in \mathbf{R}[X_1, \ldots, X_n]$ (i.e. $f(x_1, \ldots, x_n) > 0$ for all $(x_1, \ldots, x_n) \in \mathbf{R}^n$, $(x_1, \ldots, x_n) \neq (0, \ldots, 0)$) it is possible to *construct* a representation (2) for f where $g_0, \ldots, g_r$ are homogeneous polynomials.

The papers of Habicht and Kreisel raise the question whether there is an upper bound for the number r of squares, depending only on n but not on the (positive semi-definite) function $f \in R_0[X_1, \ldots, X_n]$. We will give a positive answer to this question in the next section.

It is quite clear that neither the mathematical proof by Artin and Lang nor the logical proof by Robinson can provide such a quantitative result. Therefore it will be necessary to employ other methods.

§3. Quantitative Bounds for the Number of Squares

The main result of this section will be that in $R(X_1, \ldots, X_n)$ every sum of squares can be written as a sum of at most 2^n squares whenever the coefficient field R is real-closed. This of course answers the question at the end of the last section. But it should be pointed out right at the beginning that our method does not give a new proof of Artin's theorem because it applies only to functions f which are already known to be sums of squares. In other words, we will "only" prove that the quadratic form

$$\varphi = \underbrace{\ll 1, \ldots, 1 \gg}_{n},$$

which is the unit form of dimension 2^n, represents every sum of squares in $R(X_1, \ldots, X_n)$.

The idea that 2^n is the "right" number of squares to look at comes from an unpublished paper of J. Ax on ternary definite rational functions. The essential idea of Ax was to exhibit the deep connection of this question with the theorem of Tsen and Lang on C_n-fields. In addition he employed methods from cohomology theory which were just sufficient to solve the cases $n \leq 3$ but did not go beyond.

I am very grateful to J. Ax for sending me a preprint of his manuscript during my stay in Cambridge in 1966. This was one of the most lucky occurences in my life. After a careful study of the manuscript I realized that the reduction to C_n-fields should be maintained but that the use of cohomology should be replaced by multiplicative quadratic forms.

In fact it will turn out that the following results and proofs are completely independent of sections 1 and 2 of this chapter but use some main results from Chapters 2 and 5. At the same time we will see that the field $R(X_1, \ldots, X_n)$ may be replaced by any field of transcendence degree n over R and that $\ll 1, \ldots, 1 \gg$ may be replaced by any multiplicative form $\ll a_1, \ldots, a_n \gg$, such giving a much more general result than originally intended.

We start with a lemma on multiplicative quadratic forms which is the counterpart to the result from Chapter 1, Theorem 1.8, that an arbitrary quadratic form φ which represents $a_1 \in K^\bullet$ has a diagonalization $\varphi = \langle a_1, \ldots \rangle$ (if char $K \neq 2$).

3.1 Lemma. Let K be a field with char $K \neq 2$, let $n \in \mathsf{N}$. Also let

$$\varphi = \ll a_1, \ldots, a_n \gg = \langle 1, a_1 \rangle \otimes \ldots \otimes \langle 1, a_n \rangle = \langle 1 \rangle \oplus \varphi'$$

be a regular multiplicative form of dimension 2^n over K with "pure part" φ'. (By Witt's theorem φ' is uniquely determined up to isometry.) Let $b_1 \in D_K(\varphi')$. Then there exist elements $b_2, \ldots, b_n \in K^\bullet$ such that

$$\varphi \cong \ll b_1, \ldots, b_n \gg.$$

PROOF by induction on n. $n = 1 \implies \varphi' = \langle a_1 \rangle \implies b_1 = a_1 c^2$ with $c \in K^\bullet \implies \langle a_1 \rangle \cong \langle b_1 \rangle \implies \ll a_1 \gg \cong \ll b_1 \gg$. For the inductive step

$n - 1 \to n$ (with $n \geq 2$) we note that we may now assume the lemma to hold for *all* multiplicative forms of dimension 2^{n-1}. Let φ and b_1 be as stated.

Write $\psi = \ll a_1, \ldots, a_{n-1} \gg = \langle 1 \rangle \oplus \psi'$. Then $\varphi = \psi \oplus a_n \psi$, $\varphi' = \psi' \oplus a_n \psi$, $b_1 = b_1' + a_n b$ with $b_1' \in D_K(\psi')$, $b \in D_K(\psi)$. The case $b = 0$, $b_1 = b_1'$ is trivial: By the inductive hypothesis we get

$$\psi \cong \ll b_1, \ldots, b_{n-1} \gg, \quad \varphi \cong \ll b_1, \ldots, b_{n-1}, a_n \gg$$

for suitable elements $b_2, \ldots, b_{n-1} \in K^\bullet$. Let us now assume $b \neq 0$. Then $\psi \cong b\psi$ since ψ is strictly multiplicative. Hence we may assume $b = 1$. If also $b_1' = 0$ we get $b_1 = a_n$. This case is again trivial, since $\varphi \cong \ll a_n, a_1, \ldots, a_{n-1} \gg$.

In the decisive case we have $b_1 = b_1' + a_n$ with $b_1' \in D_K(\psi')$. Here the induction hypothesis shows $\psi \cong \ll b_1', b_2, \ldots, b_{n-1} \gg$. This implies

$$\begin{aligned} \varphi &\cong \quad \ll b_1', b_2, \ldots, b_{n-1}, a_n \gg \cong \ll b_1', a_n, b_2, \ldots, b_{n-1} \gg \\ &\cong \quad \ll b_1', a_n \gg \otimes \ll b_2, \ldots, b_{n-1} \gg . \end{aligned}$$

Lemma 1.3 of Chapter 2 shows that $\langle b_1', a_n \rangle \cong \langle b_1, b_1 b_1' a_n \rangle$ since $b_1 = b_1' + a_n \neq 0$. This implies

$$\ll b_1', a_n \gg \cong \langle 1, b_1', a_n, b_1' a_n \rangle \cong \langle 1, b_1, b_1' a_n, b_1 b_1' a_n \rangle \cong \ll b_1, b_1' a_n \gg .$$

Hence

$$\varphi \cong \ll b_1, b_n \gg \otimes \ll b_2, \ldots, b_{n-1} \gg \cong \ll b_1, \ldots, b_n \gg$$

with $b_n := b_1' a_n$.

In order to formulate our main result under the weakest possible condition on the field K we introduce the following.

3.2 Definition. Let K be a field, char $K \neq 2$, let $L = K(\sqrt{-1})$. K is said to have the property (U_n) or to be a U_n-field if for every multiplicative form $\varphi = \ll a_1, \ldots, a_n \gg$ of dimension 2^n over K the induced form φ_L is universal over L ($n \in \mathbb{N}$).

With this definition we get

3.3 Theorem. Let K be a U_n-field, let $\varphi = \ll a_1, \ldots, a_n \gg$ over K, let $b \in \sum K$ be a sum of squares. Then φ represents b over K. (For the exceptional case $b = 0$ we allow the trivial representation.)

PROOF. We can assume that φ is anisotropic over K since isotropic forms are always universal. If $i = \sqrt{-1} \in K$ then $L = K$ and φ is universal over K. Therefore we may suppose $i \notin K$. Let $b = b_1^2 + \ldots + b_m^2$. We shall use induction on m. The case $m = 1$ is of course trivial since every multiplicative form φ represents 1.

1) Assume now $m = 2$, $b = b_1^2 + b_2^2$, $b_1 b_2 \neq 0$. The element

$$\beta = b_1 + b_2 i$$

generates $K(i)$ over K, i.e. we have $L = K(i) = K(\beta)$. By our assumption φ_L is universal over L. This shows that there are 2^n-dimensional vectors u, v over K such that

$$\varphi_L(u + \beta v) = \beta, \quad \text{or}$$

(1) $$\qquad \varphi(u) + 2\beta b_\varphi(u, v) + \beta^2 \varphi(v) = \beta.$$

Since $\beta \notin K$ the irreducible equation for β over K is given by

(2) $$\qquad (\beta - b_1)^2 + b_2^2 = \beta^2 - 2b_1\beta + b = 0.$$

Comparing equations (1) and (2) we find

$$\varphi(u) = b\varphi(v), \quad 2b_\varphi(u, v) - 1 = -2b_1\varphi(v).$$

As $\beta \notin K$ we must have $\varphi(v) \neq 0$, hence $b = \frac{\varphi(u)}{\varphi(v)}$. This implies $b \in D_K(\varphi)$ since φ is multiplicative over K.

2) We come to the inductive step $m \to m + 1$ $(m \geq 2)$. In other words, we assume now that the theorem holds for every m-fold sum of squares $b \in K$ and every 2^n-dimensional multiplicative form χ and we have to prove the corresponding statement for the element $c = 1^2 + b$ instead of b.

Of course we may assume $c \neq 0$. Write $\varphi = \langle 1 \rangle \oplus \varphi'$ as in Lemma 3.1. By the induction hypothesis φ represents b, i.e. we have $b = b_0^2 + b'$ with $0 \neq b' \in D_K(\varphi')$. Consider the multiplicative form $\psi = \varphi \otimes \langle 1, -c \rangle$ of dimension 2^{n+1}. We have

$$\psi = \langle 1 \rangle \oplus \varphi' \oplus (-c)\varphi = \langle 1 \rangle \oplus \psi'.$$

ψ' represents the element

$$b' - c = (b - b_0^2) - (1 + b) = -1 - b_0^2.$$

Lemma 3.1 then implies:
There exist elements $c_1, \ldots, c_n \in K^\bullet$ such that

$$\psi \cong \langle\!\langle -1 - b_0^2, c_1, \ldots, c_n \rangle\!\rangle,$$

or $\psi \cong \langle 1, -1 - b_0^2 \rangle \otimes \chi \cong \chi \oplus (-1 - b_0^2)\chi$ with

$$\chi = \langle\!\langle c_1, \ldots, c_n \rangle\!\rangle.$$

Apply the induction hypothesis again, this time to the form χ and the element $1 + b_0^2$. Conclude that χ represents $1 + b_0^2$ over K. Then $\chi \cong (1 + b_0^2)\chi$ since χ is strictly multiplicative, and $\psi \cong \varphi \oplus (-c)\varphi \sim 0$. It follows that $\varphi \cong c\varphi$ over K which in turn implies that φ represents c over K. This proves the theorem.

3.4 Corollary. Let K be a field of transcendence degree n over a real-closed field R. Then every multiplicative form $\varphi = \ll a_1, \ldots, a_n \gg$ over K represents every element $b \in \sum K$, in particular every $b \in \sum K$ is a sum of 2^n squares in K. If in addition K is nonreal then φ is universal over K.

PROOF. By Corollary 1.5 of Chapter 5 φ is universal over $K(i)$ since $C = R(i)$ is algebraically closed. Hence 3.3 applies. If K is nonreal, say $s(K) = s < \infty$, then the quadratic form $(s + 1) \times \langle 1 \rangle$ is isotropic, hence universal over K. This shows $K = \sum K$.

Notes.

(1) For the quantitative aspect of Hilbert's 17th problem we need 3.4 only for the special form $\varphi = \ll \underbrace{1, \ldots, 1}_{n} \gg = 2^n \times \langle 1 \rangle$. But an inspection of the proof of 3.3 shows that the inductive step worked only because we were considering the more general forms $\varphi = \ll a_1, \ldots, a_n \gg$. Actually all the known proofs of 3.3 and 3.4 make use of these multiplicative forms, not just of the unit form $\varphi = 2^n \times \langle 1 \rangle$!

(2) For $n = 1$ Corollary 3.4 contains the result of Witt that binary forms are universal over every nonreal field K of transcendence degree 1 over R. This comes from the fact that every binary form $\varphi = a_0\langle 1, a_1 \rangle$ is a scalar multiple of a 1-fold multiplicative form. This result has been mentioned in Chapter 5 (after 2.4).

We finish this chapter with an easy generalization of Theorem 3.3.

3.5 Theorem. Let K be a U_n-field, let $\varphi = \ll a_1, \ldots, a_n \gg$. Then the following statements hold:

(1) $D_K^{\cdot}(m \times \varphi) = D_K^{\cdot}(\varphi)$ for all $m \in \mathbf{N}$.

(2) The set $D_K^{\cdot}(\varphi) \cup \{0\}$ is closed under addition and multiplication.

(3) $D_K^{\cdot}(\varphi) \cup \{0\} = \cap P$ where P runs over all orders of K such that

$$a_1, \ldots, a_n \in P.$$

If K is nonreal or if no order P with $a_1, \ldots, a_n \in P$ exists then the (empty) intersection $\cap P$ has to be replaced by K. In this case φ is universal.

PROOF.

(1) Clearly $D_K^{\cdot}(\varphi) \subseteq D_K^{\cdot}(2 \times \varphi) \subseteq D_K^{\cdot}(3 \times \varphi) \subseteq \ldots$. Conversely, let $d \in D_K^{\cdot}(2 \times \varphi)$. Then $d = a + b$ where a and b are represented by φ and w.l.o.g. $ab \neq 0$. Write $d = a(1 + \frac{b}{a})$. Then $\frac{b}{a} \in D_K^{\cdot}(\varphi)$ since φ is multiplicative and $c = 1 + \frac{b}{a} \in D_K^{\cdot}(\varphi)$ by the proof of 3.3. Since also $a \in D_K^{\cdot}(\varphi)$ we get $d = ac \in D_K^{\cdot}(\varphi)$. This proves statement (1) for $m = 2$. Inductively, let us assume $D_K^{\cdot}((m-1) \times \varphi) = D_K^{\cdot}(\varphi)$ for some $m \geq 2$ and take an element $d \in D_K^{\cdot}(m \times \varphi)$. Then $d = a + b$ where w.l.o.g. $a \in D_K^{\cdot}(\varphi)$, $b \in D_K^{\cdot}((m-1) \times \varphi) = D_K^{\cdot}(\varphi)$. This shows $d \in D_K^{\cdot}(2 \times \varphi) = D_K^{\cdot}(\varphi)$ and proves (1).

(2) follows immediately from (1) by observing that $D_K^{\cdot}(\varphi)$ is closed under multiplication.

(3) Every element $d \in D_K^{\cdot}(\varphi) \cup \{0\}$ is a sum of elements of the form $c^2 a_1^{\nu_1} \ldots a_n^{\nu_n}$ with $c \in K$, $\nu_1, \ldots, \nu_n \in \{0, 1\}$. This shows: If $a_1, \ldots, a_n \in P$ for some order P of K then $D_K^{\cdot}(\varphi) \cup \{0\} \subseteq P$. Conversely, if $D_K^{\cdot}(\varphi) \cup \{0\} \subseteq P$ then $a_1, \ldots, a_n \in P$ since $a_i \in D_K^{\cdot}(\varphi)$.

We have the following possibilities:

- Either $-1 \notin D_K^{\cdot}(\varphi)$. Then property (2) and Definition 1.3 show that $T = D_K^{\cdot}(\varphi) \cup \{0\}$ is a preorder of K. Then our claim (3) follows from Proposition 1.7.

- Or $-1 \in D_K^{\cdot}(\varphi)$. Then $2 \times \varphi = \varphi \oplus \varphi$ is isotropic, hence universal, and $D_K^{\cdot}(\varphi) = D_K^{\cdot}(2 \times \varphi)$ shows that φ itself is also universal.

Historical Note. In [Pfister 1967$_2$] I proved that positive definite functions $f \in R(X_1, \ldots, X_n)$ are sums of 2^n squares. The generalization to Theorem 3.3 and the simplified proof given here are in [Pfister 1971$_1$ and 1971$_2$]. They are partly based on ideas of E. Witt. Theorem and proof are also given in [S, Ch. 4, Theorem 2.1]

Chapter 7

The Pythagoras Number

§1. Results for Fields

The results of the last chapter, in particular Corollary 3.4, suggest we introduce another invariant concerning sums of squares in a field or ring. It has become standard to call it the Pythagoras number. The definition is as follows:

1.1 Definition. Let A be a commutative ring with $1 \neq 0$.

(1) For $a \in \sum A$ the minimal number $n \in \mathbb{N}$ such that

$$a = a_1^2 + \ldots + a_n^2$$

with $a_i \in A$ $(i = 1, \ldots, n)$ is called the *length of a*. We denote it by $\ell(a)$.

(2) The number $p(A) := \sup\{\ell(a) : a \in \sum A\}$ is called the *Pythagoras number of A*.

1.2 Examples.

(1) $\ell(0) = \ell(1) = 1$, therefore $1 \leq p(A) \leq \infty$ for every A.

(2) $p(\mathbf{R}) = p(\mathbf{C}) = 1$.

(3) If $2 = 0$ in A then $p(A) = 1$, in particular $p(\mathbf{F}_q) = 1$ if q is a power of 2.

(4) $p(\mathbf{F}_q) = 2$ for every odd prime power q.

PROOF. (1) and (2) are trivial, (3) follows inductively from $a_1^2 + a_2^2 = (a_1 + a_2)^2$ for $a_1, a_2 \in A$, (4) follows from Example 1.10(3) of Chapter 2.

Note. A field K is called *pythagorean* if $p(K) = 1$. For such a field every triangle in the affine plane $K \times K$ which has a right angle satisfies the (rational) Pythagoras theorem: If a, b are the sides of the triangle the hypotenuse c is given by $a^2 + b^2 = c^2$, and we have $c \in K$ (c is unique up to sign).

For a nonreal ring A we have

$$\ell(-1) = s(A) < \infty$$

by the definition of the level $s(A)$ in Chapter 3. In this case the invariants $s(A)$ and $p(A)$ are intimately related. We have

1.3 Lemma. Assume $s(A) < \infty$ and $2 \in A^\bullet$. Then

$$\sum A = A \quad \text{and} \quad s(A) \le p(A) \le s(A) + 1.$$

PROOF. The inequality $s(A) = \ell(-1) \le p(A)$ is clear by the definition of $p(A)$. Take now an arbitrary element $a \in A$. Then the equation

$$a = \left(\frac{a+1}{2}\right)^2 + (-1)\left(\frac{a-1}{2}\right)^2$$

implies $a \in \sum A$, $\ell(a) \le 1 + \ell(-1) = 1 + s(A)$. This gives the upper estimate $p(A) \le s(A) + 1$.

From now on we assume that K is a field (with characteristic different from 2) leaving the more difficult case of rings for section 2.

1.4 Examples.

(1) If K is a local field then

$$p(K) = \min\{s(K) + 1,\ 4\}.$$

(2) If K is a nonreal number field then again

$$p(K) = \min\{s(K) + 1,\ 4\}.$$

(3) If K is a real number field then

$$p(K) = \begin{cases} 4 \text{ if } K \text{ has a dyadic place } \mathfrak{p} \text{ with } [K_\mathfrak{p} : \mathbf{Q}] \text{ odd,} \\ 3 \text{ otherwise.} \end{cases}$$

In particular $p(\mathbf{Q}) = 4$, $p(\mathbf{Q}(\sqrt{5})) = 3$.

(4) If K is a field of transcendence degree n over a real-closed field R then

$$p(K) \le 2^n.$$

PROOF.

(1) Like any 4-dimensional quadratic form the unit form $\langle 1, 1, 1, 1 \rangle$ is universal over K. This and Lemma 1.3 imply $s(K) \le p(K) \le \min\{s(K)+1, 4\}$. For $s(K) = 4$ this gives the desired result. For $s(K) = 1$ or 2 it is easy to see that a prime element π of K is not represented by the quadratic form $s \times \langle 1 \rangle$, hence we must have $p(K) = \ell(\pi) = s(K) + 1$.

(2) Since K is nonreal the Hasse–Minkowski theorem shows that the form $\langle 1,1,1,1 \rangle$ is universal over K. Then the proof of part (1) applies.

(3) For an element $a \in \sum K$ the 5-dimensional form

$$\varphi = \langle 1,1,1,1,-a \rangle$$

is isotropic over every completion of K including the real completions. By Hasse–Minkowski φ is isotropic over K, hence $\ell(a) \leq 4$ and $p(K) \leq 4$. [This result is sometimes referred to as the "Lagrange–Hilbert–Siegel" theorem. Compare Example 1.2 of Chapter 3]. Furthermore we get

$$p(K) = \max_{\mathfrak{p}}\{p(K_{\mathfrak{p}})\}$$

where $\mathfrak{p}$ runs through all places of K. But $p(\mathbf{R}) = p(\mathbf{C}) = 1$, hence it suffices to take the maximum over the non-archimedean places of K and to apply part (1) together with Example 1.2(6) of Chapter 3. This gives the upper estimate for $p(K)$.

For the lower estimate $p(K) \geq 3$ it remains to show that $s(K_{\mathfrak{p}}) \geq 2$ for at least one non-archimedean completion $K_{\mathfrak{p}}$ of K. This follows from the "Global Square Theorem" (see e.g. [O'M, Chapter VI, 65:15]). Otherwise -1 would be a square in almost all completions of K, hence in K. This contradicts our assumption "K real".

(4) This statement coincides with Corollary 3.4 of Chapter 6.

Question: Is the bound in Example (4) sharp for the field

$$K = \mathbf{R}(x_1,\ldots,x_n)?$$

This is trivial for $n \leq 1$.
For $n = 2$ it can be shown that the Motzkin polynomial

$$p(x_1,x_2) = 1 - 3x_1^2 x_2^2 + x_1^4 x_2^2 + x_1^2 x_2^4$$

from Chapter 1, Example 2.4 is not a sum of 3 squares in $\mathbf{R}(x_1,x_2)$, hence $p(\mathbf{R}(x_1,x_2)) = 4$. The proof can be found in [Cassels–Ellison–Pfister 1971]. It cannot be given here since it involves the theory of elliptic curves over algebraic function fields. Another proof for the fact that $p(\mathbf{R}(x_1,x_2)) = 4$ has recently been given by Colliot-Thélène [1993]. This proof involves other nontrivial things from Algebraic Geometry but no elliptic curves. At present no elementary proof for the fact that $p(K) = 4$ is known.
For $n \geq 2$ we get by induction

$$\ell(p(x_1,x_2) + x_3^2 + \ldots + x_n^2) = 4 + (n-2) = n+2,$$

hence $n + 2 \le p(\mathbf{R}(x_1, \ldots, x_n)) \le 2^n$. This is an immediate consequence of Chapter 1, Theorem 3.2.

We now turn to the question of which natural numbers m may occur as Pythagoras numbers of fields. If K is nonreal then Lemma 1.3 tells us that $p(K) = s(K)$ or $s(K) + 1$ where $s(K) = s = 2^\sigma$ is a power of 2 by Theorem 1.3 of Chapter 3. We shall show that both possibilities for $p(K)$ can actually occur for any given $\sigma \in \mathsf{N}_0$ (compare Theorem 1.4 of Chapter 3).

1.5 Proposition. Let $\sigma \in \mathsf{N}_0$, let K be a nonreal field with char $K \ne 2$ and $s(K) = 2^\sigma$. Then we have

(1) The fields $K(t)$ and $K((t))$ have Pythagoras number $2^\sigma + 1$.

(2) There exists an algebraic extension field L of K with $p(L) = 2^\sigma$.

PROOF.

(1) We have $\ell(-1) = s$ in the three fields K, $K(t)$, $K((t))$. An easy degree argument shows that the quadratic form

$$\varphi = s \times \langle 1 \rangle$$

does not represent the element t in $K(t)$ and $K((t))$ since φ is anisotropic in K. Therefore $\ell(t) > s$ in $K(t)$ and $K((t))$. On the other hand, Lemma 1.3 shows that the Pythagoras number of $K(t)$ and $K((t))$ is bounded by $s + 1$. This proves statement (1).

(2) Let $\bar{K}$ be a fixed algebraic closure of K. Consider the set $\mathcal{L}$ of all fields L, $K \subseteq L \subseteq \bar{K}$, such that $\varphi = s \times \langle 1 \rangle$ remains anisotropic over L. By inclusion $\mathcal{L}$ is inductively ordered. By Zorn's lemma there exists a *maximal* element in $\mathcal{L}$ which we denote by L again. Clearly $p(L) \ge s$ since $\ell(-1) = s$ in L. We claim that φ is *universal* over L. This of course implies $p(L) = s = 2^\sigma$.

Assume $a \in L^\bullet$, $a \notin D_L(\varphi)$. Since φ is multiplicative and $-1 \in D_L(\varphi)$ we conclude that $-a \notin D_L(\varphi)$, in particular $-a \notin L^2$. By the maximality of L the form φ is isotropic over the field $M = L(\sqrt{-a})$. Therefore we get nonzero vectors u, v with components in L such that

$$\varphi(u + v\sqrt{-a}) = 0, \quad \varphi(u) - a\varphi(v) = 0.$$

This implies $a = \frac{\varphi(u)}{\varphi(v)} \in D_L(\varphi)$, a contradiction.

From 1.3 and 1.5 it is clear that the Pythagoras number $p(K)$ is of particular interest when the field K is formally real. The examples given in 1.2 and 1.4

show that the numbers $p = 1, 2, 3, 4$ can be realized as Pythagoras numbers of real fields. In analogy with Proposition 1.5 Prestel has proved the following.

1.5' Proposition. Let $\sigma \in \mathsf{N}_0$. There exist real fields K_σ, L_σ, L_∞ with Pythagoras numbers $2^\sigma + 1$, 2^σ, ∞. In addition, these fields can be chosen to be subfields of R with a unique (archimedean) ordering. K_σ and L_σ have finite transcendence degree (depending on σ) over Q, but are not finitely generated over Q.

The main idea of the proof for K_σ is to construct two different henselian closures H_1, H_2 of a suitable function field such that H_1 is real-closed and H_2 is nonreal with $s(H_2) = 2^\sigma$. Then $K_\sigma := H_1 \cap H_2$ and $p(K_\sigma) = 2^\sigma + 1$. The proof for L_σ is essentially the same as for part (2) of Proposition 1.5. For details I refer the interested reader to the original paper [Prestel 1978].

These results immediately lead to the following question which seems to be completely open:

1.6 Open Problem. Does there exist a real field K with finite Pythagoras number such that $p(K)$ is not of the form 2^σ or $2^\sigma + 1$ for some $\sigma \in \mathsf{N}_0$, say $p(K) = 6$?

The computation of the Pythagoras number of an effectively given field will be at least as difficult as the corresponding problem for the level. Compare with the Note on page 43 in Chapter 3. Nevertheless there are some very interesting results for fields K which are generated over Q by finitely many (few) elements.

We start with a slightly more general result.

1.7 Theorem. Let φ be a strictly multiplicative quadratic form over a field K. Let

$$f(x) = c_m p_1(x) \ldots p_r(x) \in K[x]$$

be a square-free polynomial of degree $m > 0$ over K with leading coefficient $c_m \neq 0$ and monic irreducible factors p_i ($i = 1, \ldots, r$). Then the following statements are equivalent:

(1) φ represents $f(x)$ over $K(x)$.

(2) φ represents c_m over K, and φ is isotropic over the fields

$$L_i = K[x]/(p_i(x)), \quad i = 1, \ldots, r.$$

PROOF. Note that by Theorem 2.2 of Chapter 1 (which also holds for char $K = 2$) statement (1) implies the stronger statement: φ represents $f(x)$ over $K[x]$. Note also that by Theorem 3.2 of Chapter 2 (and the remarks in section 4 of that chapter concerning char $K = 2$) statement (2) implies the stronger statement: φ is hyperbolic over L_i. Clearly (1) and (2) are both satisfied if φ is isotropic over K. So we can assume φ to be anisotropic.

(1) $\Longrightarrow$ (2): Comparing highest coefficients in a representation

$$\varphi(f_1, \ldots, f_n) = f \quad \text{over } K[x]$$

we conclude that φ represents c_m over K. p_i cannot divide all the polynomials $f_1, \ldots, f_n$ since f is square-free. Let α_i be a root of p_i, hence $L_i = K(\alpha_i)$. Put $\beta_\nu = f_\nu(\alpha_i)$, $\nu = 1, \ldots, n$. Then

$$\varphi(\beta_1, \ldots, \beta_n) = f(\alpha_i) = 0 \text{ in } L_i.$$

This proves the second claim of (2) since $(\beta_1, \ldots, \beta_n) \neq (0, \ldots, 0)$. (Note that we did not use the hypothesis "φ multiplicative" for this part of the proof.)
(2) $\Longrightarrow$ (1): Since φ is multiplicative and represents c_m it is clearly enough to prove that φ represents every prime factor $p = p_i$ of f. We will now fix p and its associated field $L = K[x]/(p(x))$. The proof is by induction on $d = \deg p$. The case $d = 1$, $L = K$ is impossible, hence true!

Let $\varphi(\beta_1, \ldots, \beta_n) = 0$ with $\beta_\nu \in L = K(\alpha)$, not all $\beta_\nu = 0$. Then $\beta_\nu = g_\nu(\alpha)$, $\deg g_\nu < d$. Without loss of generality we assume $(g_1, \ldots, g_n) = 1$. Then we have

(*) $$\varphi(g_1(x), \ldots, g_n(x)) = p(x)h(x)$$

where $\deg h \leq d - 2$. If $h(x) = c_0 \neq 0$ then φ represents c_0 over K and (being multiplicative) $p(x)$ over $K(x)$, and we are done. If $h(x)$ has a non-constant normed prime factor q with root γ then $\varphi(g_1(\gamma), \ldots, g_n(\gamma)) = 0$ where $g_1(\gamma), \ldots, g_n(\gamma)$ cannot vanish simultaneously since $(g_1, \ldots, g_n) = 1$. This shows that φ is isotropic over the field $L' = K(\gamma)$. Now the induction hypothesis applies and shows that φ represents $q(x)$ over $K(x)$. This applies for every monic prime factor q of h. Finally, φ represents $h(x)$ over $K(x)$, and (*) shows that φ represents $p(x)$ over $K(x)$ since φ is multiplicative.

Note. This proof resembles very much the proof of Springer's theorem in the last chapter. In fact, the main idea of the proof comes from an old theorem of Legendre on ternary quadratic forms over $\mathbf{Z}$. For a slightly different proof see [L, Ch. X, Theorem 2.9].

1.8 Corollary. ([Landau 1906])

$$p(\mathbf{Q}(x)) \leq 8.$$

PROOF. We have to show that every positive semi-definite rational function $f(x) \in \mathbf{Q}(x)$ is a sum of eight squares. We may assume that $f(x)$ is a square-free polynomial

$$f(x) = c_m p_1(x) \ldots p_r(x) \in \mathbf{Q}[x]$$

of positive degree. Then $c_m \in \sum \mathbf{Q}$, hence c_m is a sum of four squares in $\mathbf{Q}$. The factors $p_i(x)$ of $f(x)$ must be positive definite: otherwise p_i would have a simple real zero $\alpha_i \in \mathbf{R}$ which would also be a simple zero of f since $p_1, \ldots, p_r$ are pairwise coprime. But then $f(x)$, when considered as a polynomial in $\mathbf{R}(x)$, would change sign at α_i which is impossible. Hence all roots of $p_i(x)$ are complex, in other words the field $L_i = \mathbf{Q}[x]/(p_i(x))$ is totally imaginary, i.e. nonreal in the sense of Artin–Schreier. This shows that $s(L_i) \leq 4$ by Example 1.2(7) of Chapter 3. In particular, the multiplicative form $\varphi = 8 \times \langle 1 \rangle$ is isotropic over L_i. Now Theorem 1.7 applies.

The result of Landau can be considerably strengthened and generalized as was shown in [Pourchet 1971] and [Hsia–Johnson 1974]:

1.9 Theorem. (Pourchet, Hsia-Johnson) Let K be a number field. Then

$$p(K(x)) = \begin{cases} p(K) + 1 & \text{if } K \text{ is real,} \\ s(K) + 1 & \text{if } K \text{ is nonreal.} \end{cases}$$

In all cases $p(K(x)) \leq 5$.

For a full proof I refer to [R, Chapters 17 and 18] or to the original papers. Some parts are of course easy. If $a \in \sum K$ is an element of length $\ell(a) = p(K)$ then $a + x^2 \in K(x)$ has length $p(K) + 1$ by Cassels' theorem. This gives the lower bound for $p(K(x))$. The nonreal case is also clear since then $s(K(x)) = s(K) \leq 4$. The main idea for the real case consists in a kind of elimination of the variable x and reduction to pure number theory with heavy use of approximation and the Hasse–Minkowski theorem.

Of course this theorem immediately raises the question what happens for an algebraic extension field F of $K(x)$ or for the rational function field $K(x, y)$ in two variables over K. During the last ten years there has been considerable progress in this direction. We summarize the results in a general conjecture and some details about its present status.

1.10 Conjecture. Let K be a number field, let F be a function field of transcendence degree $d \geq 1$ over K.

(1) For $d = 1$ we have $p(F) \leq 5$.

(2) For $d \geq 2$ we have $p(F) \leq 2^{d+1}$.

For $d = 1$ the estimate $p(F) \leq 7$ has been proved by Colliot-Thélène in an appendix to a paper of K. Kato [1986]. The sharper estimate $p(F) \leq 6$ is contained in an unpublished preprint of Pop [1991]. He actually proves that every element $f \in \sum F$ whose principal divisor (f) has no "real" zeros or poles is already a sum of five squares in F. So it is very likely that $p(F) \leq 5$ (compare Open Problem 1.6). The result of Kato is a local–global principle for certain cohomology groups. It belongs to 2-dimensional class field theory and uses deep results from number theory, algebraic geometry and algebraic K-theory.

Statement (2) is true modulo a generalized version of the local–global principle of Kato and modulo the Milnor conjectures (C) at the end of Chapter 2, section 3. In particular, (2) is true for $d = 2$. All this has been shown by Colliot-Thélène and Jannsen [1991].

Conjecture 1.10 also raises the following more general question: Let K be a real field with Pythagoras number $p(K) = p$. What can be said about lower or upper bounds for $p(L)$ if L is a finite extension field of K or if $L = K(X)$?

Lower bounds are known from the work of Prestel [1978] but they are nearly useless. We have:

a) Let L/K be finite. If $p = 1$ then $p(L) \geq 1$ (trivial), if $2 \leq p \leq \infty$ then $p(L) \geq 2$, and these bounds are sharp. In fact, Prestel constructs real fields K of prescribed Pythagoras number $p = 2^\sigma \geq 2$ or $p = 2^\sigma + 1$ or $p = \infty$ and a *quadratic* real extension $L = K(\sqrt{\alpha})$ such that $p(L) = 2$. This shows that "p can drop a lot in going up".

b) Let $L = K(X)$ be purely transcendental. Then $p(L) \geq p + 1$ (clear), and we have seen examples where this lower bound is sharp (say $K = \mathbf{Q}$).

Upper bounds for $p(L)$ would be even more interesting, but unfortunately nearly nothing is known so far. In particular, it is unknown whether we can have $p(K) = p < \infty$ and $p(K(X)) = \infty$. The next example shows that there is no easy estimate like $p(K(X)) \leq 2p$.

1.11 Example. Let $K = \mathbf{Q}_{\text{pyth}}$ be the *pythagorean closure* of $\mathbf{Q}$. Then $p(K) = 1$, $p(K(X)) \geq 3$.

PROOF. By definition $\mathbf{Q}_{\text{pyth}}$ is the smallest pythagorean field containing $\mathbf{Q}$ within some fixed real closure of $\mathbf{Q}$ (say the field $\mathbf{R}_a$ of all real algebraic numbers). $\mathbf{Q}_{\text{pyth}}$ can be constructed *from above* as the intersection of all fields L with $\mathbf{Q} \subset L \subseteq \mathbf{R}_a$ and $p(L) = 1$ or *from below* as the composite of extensions $L_\omega(\sqrt{a_\omega^2 + b_\omega^2})$ where ω runs through some well-ordered set starting with $L_0 = \mathbf{Q}$ and where $a_\omega, b_\omega \in L_\omega$. Clearly $K = \mathbf{Q}_{\text{pyth}}$ is contained in the quadratic closure of $\mathbf{Q}$. By Example 2.7 of Chapter 5 this implies that the field $L = K(\alpha)$ with $\alpha^4 + \alpha + 1 = 0$ has degree $[L : K] = 4$ and that L/K has no quadratic subextension. Note that the polynomial $f(X) := X^4 + X + 1$ is positive definite over $\mathbf{R}$, hence L is nonreal. Since $K(i) \not\subseteq L$ we have $s(L) \geq 2$. Thus $\varphi = \langle 1, 1 \rangle$ remains anisotropic over $L = K[X]/(f(X))$, and Theorem 1.7 implies that $f(X)$ is not a sum of two squares in $K(X)$, i.e. $\ell(f(X)) \geq 3$. (In fact it can be shown that $\ell(f(X)) = 3$ and $p(K(X)) = 3$ by use of local considerations and the Hasse–Minkowski theorem.)

The next two propositions show that at least the length of a positive definite polynomial $f(x) \in K[x]$ of fixed degree and the Pythagoras number $p(L)$ of an extension field L/K of fixed degree $[L : K]$ are bounded from above in terms of $p(K)$.

1.12 Proposition. Let K be a real field with $p = p(K) < \infty$. Let $f(x) \in K[x]$ be a polynomial of degree $2n$ which is positive semi-definite over K (in the strong sense) for all orders of K. Then $f(x)$ is a sum of $p(n + 1)$ squares in $K[x]$.

PROOF. By Theorem 2.3 of Chapter 6 $f(x)$ is a sum of squares in $K(x)$, hence in $K[x]$ (by Theorem 2.2 of Chapter 1). Since K is real this implies that $\deg f = 2n$ must be even. We will use induction on the number n. If $n = 0$ then $f(x) = a_0$ is constant with $a_0 \in \sum K$. Thus $\ell(f) = \ell(a_0) \le p$ by the definition of p. Let now $f(x) = a_{2n}x^{2n} + \ldots + a_1 x + a_0 \in \sum K[x]$ with $n > 0$, $0 \ne a_{2n} \in \sum K$, and assume that the result has already been proved for all positive semi-definite polynomials of degree $2m$ for $0 \le m < n$. The monic polynomial $\frac{f(x)}{a_{2n}}$ is a sum of squares in $K(x)$, hence in $K[x]$. Choose a sufficiently large power of 2, say $r = 2^\ell$, such that

$$(1) \qquad \frac{f(x)}{a_{2n}} = \sum_{i=1}^{r} g_i(x)^2$$

with polynomials $g_i(x) = b_i x^n + \ldots \in K[x]$ $(i = 1, \ldots, r)$ (some of the g_i may be identically zero).

Comparing highest coefficients we have

$$(2) \qquad 1 = \sum_{i=1}^{r} b_i^2.$$

Multiply equations (1) and (2) and note Corollary 2.4 of Chapter 2. This gives

$$(3) \qquad \frac{f(x)}{a_{2n}} = \left(\sum_{i=1}^{r} b_i g_i(x) \right)^2 + \text{a sum of } r - 1 \text{ squares}$$

$$= h_1(x)^2 + h(x) \text{ with } h(x) \in \sum K[x].$$

$h_1(x) = \sum_{i=1}^{r} b_i g_i(x) = \sum_{i=1}^{r}(b_i^2 x^n + \ldots)$ is a polynomial of degree n with highest coefficient $\sum_{1}^{r} b_i^2 = 1$. Comparing coefficients on both sides of (3) then implies $\deg h < 2n$. Finally we get

$$(4) \qquad f(x) = a_{2n} h_1(x)^2 + a_{2n} h(x)$$

where $a_{2n} h(x)$ is a positive semi-definite polynomial of degree $< 2n$, hence of degree $\le 2(n - 1)$. Now the induction hypothesis may be applied. It shows $\ell(a_{2n}) \le p$, $\ell(a_{2n}h) \le p \cdot n$, hence $\ell(f) \le p(n + 1)$.

Notes.

a) Example 1.11 shows that the estimate is sharp for the case $p = 1$, $n = 2$.

b) A less tricky proof of 1.12 has been shown to me by L. v. d. Dries.

1.13 Proposition. Let K be real, let L/K be a finite field extension. Then

$$p(L) \leq [L : K]p(K).$$

PROOF. Without loss of generality we may assume $p(K) < \infty$ and $L = K(\alpha)$, a simple extension. Put $[L : K] = n$. Let $\beta = \sum_{i=1}^{r} \beta_i^2 \in \sum L$, $\beta_i \in L$. Every β_i has the shape $\beta_i = g_i(\alpha)$ with a (unique) polynomial $g_i(x) \in K[x]$, $\deg g_i < n$. Put $f(x) = \sum_{i=1}^{r} g_i(x)^2$, a positive semi-definite polynomial of degree $\leq 2(n - 1)$. Then $\ell(f(x)) \leq p(K) \cdot n$ in $K[x]$ by Proposition 1.12. Substitute $x \to \alpha$, $f(x) \to \sum \beta_i^2 = \beta$. This immediately gives $\ell(\beta) \leq n \cdot p(K)$ in L. Since β was an arbitrary element of $\sum L$ this finishes the proof.

Note. Sharper estimates have been proved by Leep [1988] if K satisfies a mild extra condition or if $[L : K] = 2$.

§2. Results for Rings

In this section we collect some typical results and open problems about the Pythagoras number of rings. They go in the same direction as the corresponding questions for the level (see section 2 of Chapter 3): For rings which turn up in number theory or are "very near" to fields the Pythagoras number tends to be finite and near to the Pythagoras number of the corresponding field. But for more general rings, even integral domains, the Pythagoras number is usually infinite. This suggests we modify the definition though it is not yet obvious how to do this.

For most proofs I refer to the extensive survey article [Choi–Dai–Lam–Reznick 1982] or to other original papers since the proofs are of different a nature and do not fit into our main method of using quadratic forms.

2.1 Lemma. (Joly–Peters) Let R be a (commutative) ring with finite level $s(R)$. Then we have

(1)
$$s(R) \leq p(R) \leq s(R) + 2.$$

(2) If $2 \in R^{\bullet}$ or if $s(R)$ is even then $p(R) \leq s(R) + 1$.

PROOF.

(1) Let $a = \sum_{i=1}^{n} a_i^2 \in \sum R$. Then

$$a = (1 + \sum_{i=1}^{n} a_i)^2 - 1 - 2b = (1 + \sum a_i)^2 + b^2 - (1 + b)^2$$

where $b = \sum_{i=1}^{n} a_i + \sum_{1 \le i < j \le n} a_i a_j$. Replacing -1 by a sum of $s = s(R)$ squares in the above equation shows $\ell(a) \le s + 2$, hence $p(R) \le s + 2$.

(2) If $2 \in R^{\bullet}$ the result follows from Lemma 1.3. If s is even apply part (1) to the element $-a$ which is also in $\sum R$, say $-a = c_1^2 + c_2^2 - c_3^2$. Then

$$a - c_3^2 = -(c_1^2 + c_2^2) = (e_1^2 + \ldots + e_s^2)(c_1^2 + c_2^2).$$

By the classical formula

$$(e_{2i-1}^2 + e_{2i}^2)(c_1^2 + c_2^2) = (e_{2i-1}c_1 + e_{2i}c_2)^2 + (e_{2i-1}c_2 - e_{2i}c_1)^2$$

the right hand side is a sum of s squares in R, hence a is a sum of $s + 1$ squares.

Note. Estimate (1) is sharp for the ring $R = \mathbf{Z}[i][x]$ ($i^2 = -1$) and the element $2x = (1 + x)^2 + i^2 + (ix)^2$: Assume $2x = p^2 + q^2 = (p + iq)(p - iq)$ with $p, q \in R$. Then one factor, say $p - iq$, must be a constant polynomial $c \in \mathbf{Z}[i]$. This implies $2x = (2iq + c)c$, $2iq(0) + c = 0$, $2|c$, $4|2x$: contradiction.

2.2 Proposition. $p(K[x]) = p(K(x))$ for every field K and (one!) indeterminate x over K.

PROOF. This follows from Cassels' Theorem (Chapter 1, Theorem 2.2) for char $K \ne 2$ and is trivial for char $K = 2$ by Example 1.2 (3).

2.3 Proposition. (Colliot-Thélène and M. Kneser 1980) Let R be any valuation ring with $2 \in R^{\bullet}$ and K its quotient field. Then $p(R) = p(K)$.

PROOF. The proof is elementary, see [Choi–Lam–Reznick–Rosenberg 1980, (4.5)].

2.4 Examples. (Compare also Examples 1.4)

(1) If R is the ring of integers in a local field K then

$$p(R) \le 4.$$

(2) If R is an order (i.e. a ring with quotient field K, which is a finitely generated $\mathbf{Z}$-module) in a number field K then

a) $p(R_0) \leq 4$ if R_0 is the maximal order of K, K not totally real.

b) $p(R) \leq 5$ if K is not totally real or $[K : \mathbb{Q}] = 2$.

c) $p(R) < \infty$ if K is totally real.

PROOF.

(1) This follows from Proposition 2.3 if K is not dyadic. In the dyadic case it follows from the more general results of [Riehm 1964].

(2) a) and b) See [Peters 1974]. The proof relies on the strong approximation theorem for the spin group of the orthogonal group O_4 or O_5. This theorem does not hold if K is totally real. The bound for p is sharp: Let $R = \mathbb{Z} \cdot 1 \oplus \mathbb{Z} \cdot 2\sqrt{-2}$ in $\mathbb{Q}(\sqrt{-2})$. Then $a = 8 + 2(2\sqrt{-2})$ is a sum of five squares but not a sum of four squares in R since the system

$$\sum_1^4 (x_i^2 - 8y_i^2) = 8, \quad \sum_1^4 x_i y_i = 1$$

with $x_i, y_i \in \mathbb{Z}$ has no solution modulo 8.

(2) c) See [R. Scharlau 1980]. The finiteness of $p(R)$ results from the fact that any $a \in \sum R$ with sufficiently large absolute norm Na is a sum of five squares. By explicit construction of suitable orders R and elements $a \in \sum R$ of large length it is shown that $p(R)$ may be arbitrarily large.

In sharp contrast to the foregoing results we have

2.5 Theorem. The following real rings A have infinite Pythagoras number:

(1) $A = \mathbb{Z}[x]$.

(2) $A = K[x_1, \ldots, x_n]$, $n \geq 2$.

(3) A = real affine K-algebra of Krull dimension $d \geq 3$.

(4) (A, m) = regular local ring of dimension $d \geq 3$ with residue field $A/m = K$.

In the cases (2) – (4) K is an arbitrary real field.

PROOF. All four results are in the paper [Choi–Dai–Lam–Reznick 1982]. I repeat the (constructive) proof of (1): Let $g(x) \in \sum \mathbb{Z}[x]$ be an element of length m. We will show that $G(x) := 1 + (x - t)^2 g(x)$ has length $m + 1$ for every sufficiently large integer $t \in \mathbb{N}$.

Let $\deg g \leq 2(r - 1)$, let $N = \max\{g(i) : 1 \leq i \leq r\}$ and let t be such that

$$t > r \quad \text{and} \quad (t - r)^2 > 2 + N.$$

Assume

$$G(x) = \sum_{k=1}^{m} g_k(x)^2 \quad \text{in } \mathbf{Z}[x].$$

Clearly each g_k has degree $\leq r$. Write $g_k(x) = a_k + (x-t)h_k(x)$ where $a_k = g_k(t)$ and $h_k(x) \in \mathbf{Z}[x]$. Then

$$a_1^2 + \ldots + a_m^2 = G(t) = 1.$$

Without loss of generality we may assume $(a_1, \ldots, a_m) = (1, 0, \ldots, 0)$. Then

$$G(x) = (1 + (x-t)h_1(x))^2 + (x-t)(h_2(x)^2 + \ldots + h_m(x)^2)$$

or

$$(x-t)^2 g(x) = 2(x-t)h_1(x) + (x-t)^2(h_1(x)^2 + \ldots + h_m(x)^2).$$

This implies $(x-t)|h_1$. Writing $h_1(x) = (x-t)h(x)$ and cancelling $(x-t)^2$ we get

$$g(x) = 2h(x) + (x-t)^2 h(x)^2 + h_2(x)^2 + \ldots + h_m(x)^2.$$

Let i be an integer such that $1 \leq i \leq r$. We claim that $h(i) = 0$. In fact, if $a = |h(i)| \neq 0$ then $a \geq 1$, and evaluation of the last equation at $x = i$ gives

$$\begin{aligned} g(i) &\geq 2h(i) + (i-t)^2 h(i)^2 \geq -2a^2 + (t-i)^2 a^2 \\ &= a^2[(t-i)^2 - 2] > Na^2 \geq N \geq \max\{g(j) : 1 \leq j \leq r\}, \end{aligned}$$

which is a contradiction. This shows

$$h(x) = \prod_{i=1}^{r} (x-i)h'(x)$$

for some polynomial $h'(x) \in \mathbf{Z}[x]$ since $\mathbf{Z}[x]$ is a unique factorization domain. If $h' \neq 0$ then $\deg h \geq r$ and $\deg g \geq 2r + 2$, a contradiction. Therefore $h' = h = 0$. But then $g(x) = h_2(x)^2 + \ldots + h_m^2(x)$, i.e. $\ell(g(x)) \leq m - 1$ in $\mathbf{Z}[x]$, again a contradiction. Therefore $\ell(G(x)) = m + 1$, and by repeating the argument

$$p(\mathbf{Z}[x]) = \infty.$$

The above-mentioned paper also shows that the following "bad things" can happen which we collect now.

2.6 Examples.

(1) There exists a principal ideal domain R with real quotient field K such that $p(R) = \infty$, $p(K) < \infty$.

(2) There exists a principal ideal domain S with real quotient field K and a *unit* $u \in S$ such that

$$u \in \sum K, \quad \text{but} \quad u \notin \sum S.$$

(3) There exists an integral domain R with nonreal quotient field K such that

$$s(R) = \infty, \quad s(K) < \infty.$$

A specific example is given by ($i^2 = -1$)

$$R = \mathbf{R}[x, y]/(x^2 + y^2) \cong \mathbf{R}[x, ix], \quad K = \mathbf{C}(x).$$

Furthermore we have

2.7 Theorem. There exist integral domains with any prescribed Pythagoras number, in particular

$$p(A_n[x]) = n + 1$$

for any $n \in \mathbf{N}$ where $A_n = \mathbf{R}[x_1, \ldots, x_n]/(1 + x_1^2 + \ldots + x_n^2)$ is the ring of Chapter 3, Theorem 2.3.

PROOF. See [Dai–Lam 1984, Cor. 5.15] where it is shown that $\ell(x) = n + 1$ in $A_n[x]$. It would be interesting to know whether there also exist integral domains with any prescribed Pythagoras number and *real* quotient field.

Note. [Dai–Lam 1984, Cor. 10.6] Let $2^{r-1} < n < 2^r$. Then the quadratic form $\varphi = 2^r \times \langle 1 \rangle$ is isotropic but not hyperbolic over the ring A_n.

Theorem 2.5 shows that for most rings the definition of the Pythagoras number has to be modified if we want to get a finite invariant. I finish this section with some nice results of L. Mahé in this direction.

We consider commutative R-algebras A of finite transcendence degree d over a real-closed field R. (Elements $a_1, \ldots, a_r \in A$ are called *algebraically independent* over R if there is no polynomial $F(X_1, \ldots, X_r) \in R[X_1, \ldots, X_r]$, $F \neq 0$, such that $F(a_1, \ldots, a_r) = 0$ in A. The *transcendence degree* tr deg A of A over R is the maximal number of algebraically independent elements in A.) Typical examples of such algebras are the affine algebras $A = R[V]$ of d-dimensional varieties V over R (see the note at the end of Chapter 4 and the explanations before Theorem 2.7 in Chapter 6). In the latter case A is finitely generated over R (as a ring) and the transcendence degree of A coincides with the Krull dimension of A which in turn coincides with the dimension of V.

An element $0 \neq f \in A = R[V]$ is called *positive definite* if $f(x) > 0$ for all $x \in V(R)$. In the case where V has no real points, i.e. $V(R) = \emptyset$, this condition is empty. Then every element $0 \neq f \in A$, in particular the element $-1 \in A$, is positive definite. The following lemma is easily proved:

2.8 Lemma. Let $A = R[V]$, $0 \neq f \in A$. Then f is positive definite if and only if there exist elements $s_1, s_2 \in \sum A$ such that

$$f s_1 = 1 + s_2.$$

PROOF. See [Colliot-Thélène 1981] or [S, Ch. 3, Lemma 3.4]

This motivates the following definitions where now A is an arbitrary commutative ring with identity $1 \neq 0$:

2.9 Definition.

(1) An element $a \in A$ is called *totally positive* if there exist $s_1, s_2 \in \sum A$ such that
$$as_1 = 1 + s_2.$$
The subset of all totally positive elements of A is denoted by TA.

(2) For $a \in TA$ let
$$\ell^+(a) = \min\{n : as_1 = 1 + s_2 \text{ with } s_1, s_2 \in \textstyle\sum A,\ \ell(s_1) \leq n,\ \ell(s_2) \leq n\}$$
be the *modified length* of a.

(3) $p^+(A) := \sup\{\ell^+(a) : a \in TA\}$ is called the *modified Pythagoras number* of A.

(4) $p^\bullet(A) := \sup\{\ell^+(a) : a \in A^\bullet \cap TA\}$ is called the *unit Pythagoras number* of A.

2.10 Examples.

(1) $s = s(A) < \infty \iff 0 \in TA \iff TA = A \iff \sum A = A$. In this case we have $\ell^+(0) = \ell(-1) = s = p^+(A)$ (put $s_1 = 0$).

(2) For a field K we have $\sum A \backslash \{0\} = TA \backslash \{0\}$ and by Corollary 2.4 of Chapter 2 the invariants $p(K)$ and $p^+(K)$ are related by the equivalence
$$p(K) \leq 2^d \iff p^+(K) \leq 2^{d-1} \quad \text{(for any } d \geq 1).$$

(3) For $A = \mathbf{R}[x, ix]$ as in Example 2.6 (3) we have $x^2 \in \sum A$ but $x^2 \notin TA$, since $s(A) = \infty$. The weaker inclusion
$$A^\bullet \cap \textstyle\sum A \subseteq A^\bullet \cap TA$$
holds for every A:

Let $a = \sum_1^n a_i^2 \in A^\bullet$; then $s_1 := a^{-1} = \sum_1^n (\frac{a_i}{a})^2$ and $as_1 = 1 + s_2$ with $s_2 = 0$. In particular $\ell^+(a) \leq \ell(a)$.

(4) In general $A^\bullet \cap TA \nsubseteq A^\bullet \cap \sum A$. For example let $B = \mathbf{R}[X, Y]/(g)$ with $g(X, Y) = Y^2 - (X^3 + X)$. An easy substitution argument like that in Example 2.8 of Chapter 6 shows that no odd power of the element $x = X \bmod g$ is a sum of squares in B. Let now $A := B_x := B[\frac{1}{x}]$ be the "localization of B with respect to x". Then $x \in A^\bullet \cap TA$, but $x \notin \sum A$. See [Stengle 1979].

The main results of Mahé read as follows:

2.11 Theorem. Let A be a commutative algebra of transcendence degree d over a real-closed field. Then

(1)

$$
\begin{aligned}
p^+(A) &\leq 2^{d+1} - 1 &&\text{for } d \leq 3 \\
p^+(A) &\leq 2^{d+1} + d - 4 &&\text{for } d \geq 4
\end{aligned}
$$

(2)

$$
\begin{aligned}
p^\bullet(A) &\leq 2^d &&\text{for } d \leq 3, \\
p^\bullet(A) &\leq 2^d + d - 5 &&\text{for } d \geq 4.
\end{aligned}
$$

PROOF. See [Mahé 1990, 1992]. The slightly different estimates for $d \leq 3$ are due to the composition formulas for 2-, 4- and 8-fold sums of squares over A. Compare the end of section 2 in Chapter 2.

Under additional assumptions on A the estimates can be improved. For instance we have

2.12 Proposition.

(1) Let A be a semi-local R-algebra of transcendence degree d (i.e. A has only finitely many maximal ideals). Then the unit Pythagoras number satisfies

$$p^\bullet(A) \leq 2^d.$$

(2) Let V be an R-variety of dimension d with affine algebra $B = R[V]$. Let $S = \{1 + b : b \in \sum B\}$, put $A = B[S^{-1}]$. A is called the *ring of regular functions* on V. Then

$$p^+(A) \leq 2^d.$$

PROOF. See [Mahé, op. cit.].

Chapter 8

The u-invariant

§1. Definitions and Examples

Throughout this chapter K will be a field. In section 1-3 we assume char $K \neq 2$. The u-invariant $u(K)$ is one of the most interesting and intricate invariants of K related to quadratic forms over K. Many natural questions about its properties are still open, let alone the possibility of computing $u(K)$ for a wide class of fields K.

We start with Kaplansky's definiton of $u(K)$ for nonreal fields.

1.1 Definition. [Kaplansky 1953] Let K be nonreal. Then

$$u = u(K) := \sup\{\dim\varphi : \varphi \text{ anisotropic quadratic form over } K\}.$$

The letter u is explained by the fact that in the case $u(K) < \infty$ every φ with $\dim\varphi > u$ is isotropic over K which implies that *every u-dimensional quadratic form over K is universal.* (This does not exclude the possibility that there may exist anisotropic universal forms ψ with $\dim\psi < u$. But we will not consider such examples here.) Clearly $u(K) \geq s(K)$ for every nonreal field since $\varphi = s \times \langle 1 \rangle$ is anisotropic over K by the definiton of the level $s = s(K)$.

1.2 Examples.

(1) $u(K) = 1$ iff K is quadratically closed. In particular $u(\mathbf{C}) = 1$.

(2) $K = \mathbf{F}_q$ finite field (q odd) $\implies u(K) = 2$.

(3) K local field (char $K \neq 2$) $\implies u(K) = 4$.

(4) More generally: Let K be complete with respect to a discrete valuation $v : K^\bullet \to \mathbf{Z}$ and with nonreal residue field k, char $k \neq 2$. Then $u(K) = 2u(k)$.

(5) K nonreal number field $\implies u(K) = 4$.

(6) K C_n-field or C_n^p-field for an odd prime number $p \implies u(K) \leq 2^n$. In particular: $u(\mathbf{C}(x_1, \ldots, x_n)) = 2^n$.

PROOF.

(1) is trivial.

(2) follows for instance from Chevalley's Theorem 2.1 in Chapter 5. For a more classical proof compare Example 1.10 (3c) in Chapter 2.

(3) follows from (4).

(4) This is a well-known theorem of T.A. Springer [1955] which also holds for char $k = 2$. In our case char $k \neq 2$ the proof is very easy, as follows: Any regular n-dimensional quadratic form φ over K is equivalent to a diagonal form

$$\langle u_1, \ldots, u_m \rangle \oplus \pi \langle u_{m+1}, \ldots, u_n \rangle$$

where π is a fixed prime element of K (i.e. $v(\pi) = 1$), where the u_i are units (i.e. $v(u_i) = 0$) and $0 \leq m \leq n$. Since char $k \neq 2$ Hensel's Lemma immediately implies

$$\langle u_1, \ldots, u_m \rangle \text{ isotropic over } K \iff \langle \bar{u}_1, \ldots, \bar{u}_m \rangle \text{ isotropic over } k,$$

where $\bar{u}_i$ denotes the image of u_i in the residue field k. If now φ is anisotropic over K then we must have $m \leq u(k)$, $n - m \leq u(k)$, hence $n = \dim \varphi \leq 2u(k)$. Conversely, if $m = u(k)$ and $\langle \bar{u}_1, \ldots, \bar{u}_m \rangle = \bar{\psi}$ is anisotropic over k let $\psi = \langle u_1, \ldots, u_m \rangle$ where $u_i \in K$ is a lift of $\bar{u}_i \in k$. Then every nonzero element a which is represented by ψ has even value $v(a)$. Since ψ is anisotropic this implies that $\varphi = \psi \oplus \pi \psi$ is anisotropic, too. This gives $u(K) \geq 2u(k)$ and finishes the proof.

(5) Use (1) and (3) and apply the Hasse–Minkowski theorem.

(6) $u(K) \leq 2^n$ is clear from the definiton of the properties C_n and C_n^p. By induction on n it is easily seen that the form

$$\ll x_1, \ldots, x_n \gg \: = \: \ll x_1, \ldots, x_{n-1} \gg \: \oplus \: x_n \ll x_1, \ldots, x_{n-1} \gg$$

is anisotropic over $C(x_1, \ldots, x_n)$: Compare degrees w.r.t. x_n. Hence $u(C(x_1, \ldots, x_n)) = 2^n$.

Example (6) shows that every power of 2 (including 2^∞) is the u-invariant of a suitable nonreal field. Even without Tsen–Lang theory we conclude from Example (4) that the iterated power series field $C((x_1)) \ldots ((x_n))$ also has u-invariant 2^n for $0 \leq n \leq \infty$. This led Kaplansky to the question whether $u(K)$ is perhaps a power of 2 for every nonreal field K. We shall see in section 3 that this is not the case. For the moment we are satisfied with proving

1.3 Proposition. $u(K) \neq 3, 5, 7$ for every nonreal field K.

PROOF.

(1) Assume $u(K) < 4$. Then every 2-fold multiplicative form $\ll a,b \gg = \langle 1,a \rangle \otimes \langle 1,b \rangle$ is isotropic, hence hyperbolic by Theorem 3.2 of Chapter 2. This implies $\langle 1,a \rangle \cong -b\langle 1,a \rangle$, i.e. $\langle 1,a \rangle$ represents $-b$, i.e. $\langle 1,a,b \rangle$ is isotropic. By scaling it follows that every ternary form $\langle a,b,c \rangle$ is isotropic over K, hence $u(K) \leq 2$.

(2) Assume now $u(K) < 8$. Then $\ll a,b,-c \gg$ is hyperbolic over K for all $a,b,c \in K^\bullet$. This implies $\ll a,b \gg \cong c \ll a,b \gg$. By Observation 3.7 of Chapter 2 every form $\tilde{\psi} \in I^2$ is a sum of forms $\ll \widetilde{a_i,b_i} \gg$, hence we have $\tilde{\psi} = c\tilde{\psi}$ in $W(K)$. Since also $H = cH$ for the hyperbolic plane $H = \langle 1,-1 \rangle$ Witt cancellation implies $\psi \cong c\psi$ for every individual quadratic form ψ with $\tilde{\psi} \in I^2$ and every $c \in K^\bullet$. In particular, ψ is universal over K.

(3) Assume now $u(K) = 5$ or 7 and let φ be an anisotropic quadratic form of dimension u. Then φ is universal. In particular, φ represents its discriminant $d = d(\varphi)$. Therefore $\varphi \cong \psi \oplus \langle d \rangle$ where ψ has dimension 4 or 6 and discriminant $d(\psi) = 1$. Thus $\tilde{\psi} \in I$ and $d(\tilde{\psi}) = 1$ which gives $\tilde{\psi} \in I^2$ by Proposition 3.6 of Chapter 2. From part (2) of the proof it follows that ψ is universal which in turn implies φ isotropic over K: contradiction.

Let us now turn to a formally real field K. Then a regular quadratic form $\varphi = \langle a_1, \ldots, a_n \rangle$ is clearly anisotropic if there is at least one ordering $>$ on K such that all a_i have the same sign with respect to $>$. In particular, $\varphi = n \times \langle 1 \rangle$ is anisotropic for all $n \in \mathbf{N}$. The naive definition of the u-invariant would give $u(K) = \infty$. In order that the dimension of anisotropic forms is possibly bounded we must therefore necessarily restrict to forms φ which are *totally indefinite*, i.e. for every order $>$ on K at least one coefficient a_i of φ is positive and at least one coefficient a_j is negative $(1 \leq i, j \leq n)$. The indices i, j may of course depend on the chosen order. However, the following example shows that this condition is not yet enough for making the dimension of φ bounded, even for very simple fields.

1.4 Example. Let $K = \mathbf{R}(x,y)$ or $\mathbf{R}((x))((y))$. Consider

$$\varphi_n = \langle \underbrace{1, \ldots, 1}_{n}, x, y, -xy \rangle,$$

where $n \in \mathbf{N}$ is arbitrary. Then φ_n is anisotropic and totally indefinite over K, $\dim \varphi_n = n + 3$.

PROOF.

(1) In every order of K at least one of the three elements $x, y, -xy$ is negative. This immediately shows that φ is totally indefinite.

(2) The form $\langle \underbrace{1,\dots,1}_{n},y \rangle$ is anisotropic over K: otherwise we would find "polynomials" resp. "power series"

$$f_i \in \mathbf{R}(x)[y], \quad \text{resp.} \quad \mathbf{R}((x))[[y]]$$

$(i = 0,\dots,n$, not all $f_i = 0)$, such that

(*) $$f_1^2 + \dots + f_n^2 = -y f_0^2.$$

The coefficients of the f_i lie in the fields $\mathbf{R}(x)$, resp. $\mathbf{R}((x))$, which are clearly real. Look now at the terms of lowest degree in y. Then the left hand side of (*) has lowest term of even degree while the right hand side of (*) has lowest term of odd degree, unless $f_0 = 0$ and $f_1^2 + \dots + f_n^2 = 0$. But this would imply $f_1 = \dots = f_n = 0$, a contradiction.

Similarly, the forms $\langle 1,\dots,1,x \rangle$ and $\langle 1,-x \rangle$ are anisotropic over $\mathbf{R}(x)$, resp. $\mathbf{R}((x))$, and also over K.

(3) Assume now that φ_n is isotropic. Then there are "polynomials", resp. "power series", $g_1,\dots,g_{n+1}$ and h_1, h_2 with respect to the variable y, not all 0, such that

(**) $$g_1^2 + \dots + g_n^2 + x g_{n+1}^2 = -y(h_1^2 - x h_2^2).$$

Again we can compare the lowest non-vanishing terms w.r.t. y. This gives $g_1^2 + \dots + g_n^2 + x g_{n+1}^2 = 0$ and $h_1^2 - x h_2^2 = 0$ in $\mathbf{R}(x)$, resp. $\mathbf{R}((x))$, from which we get $g_1 = \dots = g_{n+1} = h_1 = h_2 = 0$ by the last sentence of part (2), a final contradiction.

Example 1.4 and similar examples motivate us to look only at quadratic forms φ of *total signature* 0 when defining the u-invariant. Here for any form $\varphi = \langle a_1,\dots,a_n \rangle$ over K and any order $P \subset K$ the signature of φ w.r.t. P is defined by

$$\text{sign}_P \varphi = \sum_{i=1}^{n} \text{sign}_P a_i \in \mathbf{Z}.$$

Equivalently we could use the imbedding $K \subset K_P$ where K_P is a real closure of (K,P) and the fact that $W(K_P) = \mathbf{Z}$ (Compare Example 1.10(2) in Chapter 2). Then $\text{sign}_P \varphi = \varphi \otimes K_P \in W(K_P) = \mathbf{Z}$. By definition φ has total signature 0 iff $\text{sign}_P \varphi = 0$ for all orders P of K.

A major theorem in the algebraic theory of quadratic forms tells us that the statement "φ has total signature 0" is equivalent to the statement "The Witt class $\tilde{\varphi}$ of φ belongs to the torsion subgroup $W_t(K)$ of the Witt group $W(K)$". This was first proved in [Pfister 1966]. Better proofs can be found in the books of Lam [L, Ch. 8, Thm. 4.1] and Scharlau [S, Ch. 3, Thm. 6.2].

After these preliminaries we arrive at the approved and generally accepted definition of the u-invariant:

1.5 Definition [Elman–Lam 1973$_1$] For every field K of characteristic different from 2 the number

$$u \;=\; u(K)$$
$$:= \sup\{\dim\varphi : \varphi \text{ anisotropic quadratic form over } K \text{ with } \tilde\varphi \in W_t(K)\}$$

is called the *u-invariant of K*.

Note that this definition coincides with 1.1 if K is nonreal since then $W_t(K) = W(K)$. More precisely: $(2s) \times \varphi \sim 0$ for every form φ where s is the level of K.

1.6 Note. For a real field K the u-invariant $u(K)$ is an even number or $u(K) = \infty$: Assume $u = u(K) < \infty$ and let φ be an anisotropic torsion form with $\dim\varphi = u$, say $m \times \varphi \sim 0$ ($m \in \mathbf{N}$). Let P be an order of K. Then $\mathrm{sign}_P(m \times \varphi) = m\,\mathrm{sign}_P\varphi = 0$, hence $\mathrm{sign}_P\varphi = 0$. This implies $u = \dim\varphi \in 2\mathbf{N}_0$ since $\dim\varphi \equiv \mathrm{sign}_P\varphi \bmod 2$.

1.7 Examples (real fields).

(1) $u(K) = 0 \iff K$ is real pythagorean.
   In particular, $u(R) = 0$ for every real-closed field R.

(2) $K = k((x)) \implies u(K) = 2u(k)$.

(3) K real number field $\implies u(K) = 4$.

(4) If K is real with Pythagoras number $p(K) > 2^n$ for some $n \in \mathbf{N}_0$ then $u(K) \geq 2^{n+1}$. In particular we have $u(K) \geq p(K)$ unless $p(K) = 1$.

(5) Let K be such that $u(K(i)) \leq 2^n$. Then $u(K) < 4 \cdot 2^n$.
   In particular $u(\mathbf{R}(x_1,\ldots,x_n)) < 4 \cdot 2^n$ since $\mathbf{C}(x_1,\ldots,x_n)$ has u-invariant 2^n.

PROOF.

(1) By Definition 1.5 we have $u(K) = 0 \iff W_t(K) = 0$. On the other hand take a totally positive element $a \in \sum K$ of length $\ell(a) = m$ and take any n with $2^n \geq m$. Then $2^n \times \langle 1, -a\rangle$ is isotropic, hence hyperbolic. This shows $\langle 1, \widetilde{-a}\rangle \in W_t(K)$ for every $a \in \sum K$. Therefore $W_t(K) = 0 \iff \ell(a) = 1$ for all $a \in \sum K \iff p(K) = 1$.

(2) Let φ be an anisotropic torsion form of dimension $u(k)$ over k. Then $\varphi \oplus x\varphi$ is an anisotropic torsion form of dimension $2u(k)$ over $K = k((x))$. This shows $u(K) \geq 2u(k)$. Conversely, let ψ be anisotropic over K. Up

to squares every element of $K^{\bullet}$ is of the form a or ax with $a \in kdot$. Therefore $\psi \cong \varphi_1 \oplus x\varphi_2$ with forms φ_1, φ_2 over k. Every order P of k can be extended in exactly two ways to K making $\pm x$ positive and infinitely small, since $1 + \sum_{i=1}^{\infty} a_i x^i$ $(a_i \in k)$ is a square in K. Let P_+, P_- denote these two orders of K. Then $\text{sign}_{P_{+,-}} \psi = \text{sign}_P \varphi_1 \pm \text{sign}_P \varphi_2$. If now $\tilde{\psi} \in W_t(K)$ then $\text{sign}_{P_{+,-}} \psi = 0$ for all P, hence $\text{sign}_P \varphi_1 = \text{sign}_P \varphi_2 = 0$ for all P. This implies $\dim \varphi_{1,2} \leq u(k)$ and $u(K) \leq 2u(k)$.

(3) By Example 1.4(3) of Chapter 7 we have $3 \leq p(K) \leq 4$. Let $a \in \sum K$, $\ell(a) \geq 3$. Then $2 \times \langle 1, -a \rangle$ is an anisotropic torsion form of dimension 4. Therefore $u(K) \geq 4$. On the other hand consider a quadratic form φ over K which has total signature 0 and $\dim \varphi \geq 5$. (It would be enough if φ were totally indefinite.) Then φ is isotropic over all completions of K (including the finitely many real completions), thus φ is isotropic over K by Hasse–Minkowski. Therefore $u(K) \leq 4$.

(4) Let $a \in \sum K$, $\ell(a) > 2^n$. Then $\varphi = 2^n \times \langle 1, -a \rangle$ is an anisotropic torsion form over K (since φ is multiplicative). This shows $u(K) \geq 2^{n+1}$.

(5) See the next section where we prove a slightly stronger result.

More results and a number of open problems about the u-invariant will be given in the next two sections.

§2. The Filtration of Elman and Lam

By Theorem 3.4 of Chapter 2 the exact order of a quadratic torsion form in the Witt group $W(K)$ is a power of 2. This allows us to introduce the following finer u-invariants $u^{(i)}$:

2.1 Definition. For each $i \in \mathbb{N}_0$ let

$$u^{(i)} = u^{(i)}(K)$$
$$:= \sup\{\dim \varphi : \varphi \text{ anisotropic quadratic form over } K, 2^i \times \varphi \sim 0\}.$$

We often write u' instead of $u^{(1)}$.

It is immediately clear from this definition that $0 = u^{(0)} \leq u^{(1)} \leq \ldots$, $u = \sup_i u^{(i)}$.

2.2 Lemma.
$$u(K) = 0 \Longleftrightarrow u'(K) = 0.$$

PROOF. $\Longrightarrow$ is clear. Assume now $u(K) > 0$. Then we have at least one anisotropic form φ with $\dim \varphi \geq 1$ and $2^i \times \varphi \sim 0$ for some exponent $i = i(\varphi) >$

0. We may assume that i is minimal, i.e. $2^{i-1} \times \varphi \not\sim 0$. Let ψ be the anisotropic kernel of $2^{i-1} \times \varphi$. Then $\dim \psi \geq 1$ and $2 \times \psi \sim 2 \times (2^{i-1} \times \varphi) = 2^i \times \varphi \sim 0$. This shows $u'(K) = u^{(1)}(K) \geq \dim \psi > 0$.

Since fields with $u(K) = 0$ have been completely determined by Example 1.7 (1), we can now assume $u'(K) > 0$ if necessary.

2.3 Proposition. Let φ be a quadratic form with $\dim \varphi \geq 2$ such that $2 \times \varphi$ is isotropic. Then φ contains a binary subform ψ with $2 \times \psi \sim 0$.

PROOF. If φ itself is isotropic take $\psi = \langle 1, -1 \rangle$. If $s(K) = 1$ take any binary subform ψ of φ. Then $2 \times \psi \sim 0$. Let now φ be anisotropic and $s(K) \neq 1$. Since $2 \times \varphi$ is isotropic we find vectors u, v, both $\neq 0$, such that $\varphi(u) + \varphi(v) = 0$. Then $\varphi(u) = -\varphi(v) \neq 0$. Put $a = \varphi(u)$. Then $\varphi \cong \langle a \rangle \oplus \varphi'$ with $\dim \varphi' \geq 1$. Write

$$-a = \varphi(v) = av_1^2 + \varphi'(w).$$

Since $-1 \notin K^2$ we have $b := \varphi'(w) \neq 0$. This implies

$$\varphi \cong \langle a, b, \ldots \rangle.$$

Let $\psi := \langle a, b \rangle$. We get

$$b = -a(1 + v_1^2), \quad \psi = a\langle 1, -(1 + v_1^2) \rangle, \quad 2 \times \psi \sim 0.$$

This finishes the proof.

2.4 Theorem. [Elman–Lam 1973$_1$]. Let $0 < u' < \infty$. Then

(1)
$$u^{(i+1)} \leq u' + \frac{1}{2} u^{(i)} \text{ for } i \geq 0,$$

(2)
$$u^{(i+1)} \leq \left(\sum_{j=0}^{i} \frac{1}{2^j} \right) u' < 2u' \text{ for } i \geq 0,$$

(3)
$$u' \leq u \leq 2u' - 1.$$

PROOF.

(1) Let φ be anisotropic with $2^{(i+1)} \times \varphi \sim 0$. Whether $2 \times \varphi$ is isotropic or not we can apply the last proposition as often as possible ($r \geq 0$ times) to get

$$\varphi \cong \psi_1 \oplus \ldots \oplus \psi_r \oplus \varphi_0$$

where the ψ_ϱ are binary forms with $2 \times \psi_\varrho \sim 0$ and where either $2 \times \varphi_0$ is anisotropic (this is also true in the case where $\varphi_0 = \emptyset$ is the empty form!)

or $\dim \varphi_0 = 1$. Then $\psi = \psi_1 \oplus \ldots \oplus \psi_r$ is an anisotropic subform of φ with $2 \times \psi \sim 0$. Therefore $\dim \psi \leq u^{(1)} = u'$. Secondly $2^{(i+1)} \times \varphi \sim 0$ and $2^{(i+1)} \times \psi \sim 0$ implies

$$2^{(i+1)} \times \varphi_0 = 2^{(i)} \times (2 \times \varphi_0) \sim 0.$$

If $2 \times \varphi_0$ is anisotropic it follows that $\dim(2 \times \varphi_0) = 2 \dim \varphi_0 \leq u^{(i)}$ by the definition of $u^{(i)}$. Combining these two estimates we get

$$\dim \varphi = \dim \psi + \dim \varphi_0 \leq u^{(1)} + \frac{1}{2} u^{(i)}, \text{ i.e. (1)}.$$

If $2 \times \varphi_0$ is isotropic we must be in the exceptional case $\dim \varphi_0 = 1$ and $2 \times \varphi_0 \sim 0$ which implies $s(k) = 1$ and $2 \times \chi \sim 0$ for every quadratic form χ, hence $u^{(1)} = u^{(2)} = \ldots = u$. Then (1) is trivially true.

(2) follows from (1) by an easy induction on i.

(3) follows from (2): $u = \sup u^{(i)} \leq 2u'$ is finite (by the assumption $u' < \infty$). This implies $u = u^{(i)}$ for all $i \geq i_0$ and suitable $i_0 \in \mathbf{N}$ since the $u^{(i)}$ are integers. Finally (2) gives $u \leq 2u' - 1$.

This theorem shows that we might replace the field invariant $u(K)$ by the invariant $u'(K)$ if we are only interested in the finiteness of $u(K)$. But of course one would like to know more. We ask

2.5 Open Question. $u(K) = u'(K)$ for all fields K?
No counterexamples are known up to now. In some very special cases we can prove equality. Let I denote the fundamental ideal of $W = W(K)$.

2.6 Proposition. Suppose I^3 torsion-free, i.e. $I^3 \cap W_t = 0$, and $1 < u < \infty$. Then we have

a) u is even.

b) If $u > 2$ then there exists an anisotropic form ψ with $\dim \psi = u$ and $\tilde{\psi} \in I^2 \cap W_t$.

c) $u = u'$.

PROOF.

a) Assume that u is odd. Then K must be nonreal. Let φ be anisotropic, $\dim \varphi = u$. As in the proof of 1.3 we see that φ represents its discriminant $d = d(\varphi)$. This implies $\varphi \cong \psi \oplus \langle d \rangle$ with $\dim \psi = u - 1$ and $d(\psi) = 1$, hence $\tilde{\psi} \in I^2$. Then $\psi \cong c\psi$ for every $c \in K^\bullet$ since $\tilde{\psi} \otimes \langle 1, -c \rangle \in I^3 = I^3 \cap W_t = 0$. In particular, ψ is universal and φ is isotropic: contradiction.

b) Take φ anisotropic, $\dim \varphi = u$, $\tilde{\varphi} \in W_t$. Since now u is even we have
$d = d(\varphi) = (-1)^{u/2} \det \varphi$ and $\text{sign}(\det \varphi) = (-1)^{u/2}$ for all orderings P
of K (if any). This implies that d is totally positive and that $\langle 1, -d \rangle \in$
W_t. We consider the $(u+2)$-dimensional form $\chi = \varphi \oplus \langle 1, -d \rangle$ with
$\tilde{\chi} \in W_t$ and $d(\chi) = 1$. By the definition of the u-invariant χ is clearly
isotropic, say $\chi = \psi \oplus \langle 1, -1 \rangle$. Furthermore, we have $\tilde{\psi} \in I^2 \cap W_t$. We
claim that ψ must be anisotropic. Otherwise we get $\psi = \sigma \oplus \langle 1, -1 \rangle$,
$\chi = \sigma \oplus 2 \times \langle 1, -1 \rangle = \sigma \oplus \langle d, -1 \rangle \oplus \langle 1, -d \rangle = \varphi \oplus \langle 1, -d \rangle$. Witt's
cancellation theorem implies that $\varphi = \sigma \oplus \langle d, -1 \rangle$ with $\dim \sigma = u - 2$
and $\tilde{\sigma} = \tilde{\psi} \in I^2 \cap W_t$. As before this implies σ universal and φ isotropic,
a contradiction.

c) Since $1 < u < \infty$ and $u' \le u < 2u'$ we have $u' \ge 2$. For $u' = 2$ we
find $u < 4$, hence $u \le 2$ by 1.3, hence $u = u'$. For $u' > 2$ part b)
applies and yields a form ψ with the corresponding properties. They
imply $2 \times \tilde{\psi} \in I^3 \cap W_t = 0$ which means that $\tilde{\psi}$ is automatically of order
2. Thus we get $u' \ge \dim \psi = u$, hence $u' = u$.

A well-known theorem in the algebraic theory of quadratic forms states the
following:

If $\varphi \ne \emptyset$ is anisotropic with $\tilde{\varphi} \in I^n$ for some $n \in \mathbb{N}$ then $\dim \varphi \ge 2^n$. For a
proof see [L, Ch. 10, Thm. 3.1] or [S, Ch. 4, Thm. 5.6].

Taking this result for granted we can show:

2.7 Proposition. If $u' < 8$ then $u' = u$.

PROOF. By 2.2 and 2.4 the cases $u' = 0$, $u' = 1$ are trivial. The cases
$u' = 2$ and $I^3 \cap W_t = 0$ are covered by the last proposition, 2.6. Let us now
assume $I^3 \cap W_t \ne 0$. We shall see that this is impossible: Take any anisotropic
$\varphi \ne \emptyset$ with $\tilde{\varphi} \in I^3 \cap W_t$. Then $\dim \varphi \ge 8$.

a) $2 \times \varphi \sim 0$: This implies $u' \ge 8$, a contradiction.

b) $2 \times \varphi \nsim 0$: Let ψ be the anisotropic kernel of $2 \times \varphi$. Then $\psi \ne \emptyset$ and
$\dim \psi \ge 16$ since $\tilde{\psi} = 2 \widetilde{\times \varphi} \in I^4 \cap W_t$. This implies $u \ge 16$ whereas
$u \le 2u' - 1 \le 13$ by 2.4, again a contradiction.

The last argument can be generalized from the exponent 3 to any exponent
n. Then we get

2.8 Proposition. If $u' < 2^n$ then $I^n \cap W_t = 0$.

We shall see in the next section that the converse of this result does not
hold for $n \ge 3$.

A considerable advantage of the invariant u' instead of the invariant u is
expressed in the next proposition:

2.9 Proposition. Let φ be a quadratic form with $2 \times \varphi \sim 0$. Then $\varphi \cong \varphi_1 \oplus \ldots \oplus \varphi_r$ with $\dim \varphi_i \leq 2$ and $2 \times \varphi_i \sim 0$ $(i = 1, \ldots, r)$. For $\dim \varphi_i = 2$ we have $\varphi_i \cong a_i \langle 1, -(1 + v_i^2) \rangle$ where $a_i \in K^\bullet$, $v_i \in K$. The case $\dim \varphi$ odd can only occur if K is nonreal and $s(K) = 1$.

PROOF. For $\dim \varphi > 0$ the assumption $2 \times \varphi \sim 0$ implies that $2 \times \varphi$ is isotropic. If $\dim \varphi = 1$, $\varphi = \langle a \rangle$, then $\langle a, a \rangle \sim 0$ which gives $s = 1$. If $\dim \varphi \geq 2$ then Proposition 2.3 applies and yields $\varphi \cong \varphi_1 \oplus \chi$ where $\dim \varphi_1 = 2$, $2 \times \varphi_1 \sim 0$ and where φ_1 has the form $\varphi_1 \cong a_1 \langle 1, -(1 + v_1^2) \rangle$. Furthermore $2 \times \varphi \sim 2 \times \varphi_1 \sim 0$ imply $2 \times \chi \sim 0$. The result follows by induction on $\dim \varphi$.

2.10 Corollary. $s(K) \neq 1 \Longrightarrow u'(K)$ is an even number or ∞.

Proposition 2.9 cannot be generalized to forms of finite order > 2 in any reasonable manner. This is shown by the following.

2.11 Example. Let $K = \mathbf{Q}(x, y)$, $\varphi = \langle 1, x \rangle \otimes \langle 1, -(1 + 3x + y^2) \rangle$. Then $4 \times \varphi \sim 0$ but φ does not contain a binary subform ψ with $\hat{\psi} \in W_t(K)$.

PROOF. See [Arason–Pfister 1977].

With the help of the u'-invariant we can now prove the following theorem which immediately implies Example 1.7(5) since $u(K) \leq 2u'(K)$ for all fields K.

2.12 Theorem. Let K be a field with $s(K) \neq 1$, let $i = \sqrt{-1}$ and let $L = K(i)$. Then

$$u'(K) \leq 2(u(L) - 1) < 2u(L).$$

In particular:

$$u(L) = 1 \implies u(K) = u'(K) = 0,$$
$$u(L) = 2 \implies u(K) = u'(K) \leq 2,$$
$$u(L) = 4 \implies u(K) = u'(K) \leq 6.$$

PROOF. By assumption we have $-1 \notin K^2$, hence $[L : K] = 2$. Let φ be an anisotropic quadratic form over K with $2 \times \varphi \sim 0$. From 2.9 we conclude that

$$\varphi \cong \varphi_1 \oplus \ldots \oplus \varphi_r$$

with $r \geq 0$, $\varphi_\varrho \cong a_\varrho \langle 1, -t_\varrho \rangle$ where $a_\varrho \in K^\bullet$, $t_\varrho = b_\varrho^2 + c_\varrho^2 \in K^\bullet$. In particular, we have $\dim \varphi = 2r$.

For the moment we omit the suffix ϱ and consider a binary form $\langle 1, -t \rangle$ over K with $t = b^2 + c^2$, a sum of two squares. Consider the 1-dimensional form $\chi = \langle b + ic \rangle$ over L. For $\xi = x + iy \in L$ $(x, y \in K)$ we have

$$\chi(\xi) = (b + ic)(x + iy)^2 = [b(x^2 - y^2) - 2cxy] + i[c(x^2 - y^2) + 2bxy].$$

Thus the "imaginary part" of $\chi(\xi)$ is a binary quadratic form over K, namely

$$\operatorname{Im}\chi(\xi) = c(x^2 - y^2) + 2bxy = (x,y)\begin{pmatrix} c & b \\ b & -c \end{pmatrix}\begin{pmatrix} x \\ y \end{pmatrix} =: \psi(x,y).$$

If we diagonalize ψ we get $\psi \cong c\langle 1, -(b^2 + c^2)\rangle$. (For $c = 0$ this means $\psi \cong H$, the hyperbolic plane.)

Define now $\chi_\varrho := a_\varrho c_\varrho\langle b_\varrho + ic_\varrho\rangle$, $\varrho = 1,\ldots,r$. Then $\operatorname{Im}\chi_\varrho \cong a_\varrho c_\varrho c_\varrho\langle 1, -(b_\varrho^2 + c_\varrho^2)\rangle \cong a_\varrho\langle 1, -t_\varrho\rangle \cong \varphi_\varrho$. Finally, let $\chi := \chi_1 \oplus \ldots \oplus \chi_r$. This is an r-dimensional quadratic form over L with $\operatorname{Im}\chi \cong \varphi_1 \oplus \ldots \oplus \varphi_r \cong \varphi$ over K.

We have to show that $r \leq u(L) - 1$. Then $\dim\varphi \leq 2(u(L) - 1)$ for all anisotropic forms φ over K with $2 \times \varphi \sim 0$, hence $u'(K) \leq 2(u(L) - 1)$ by the definition of u'.

Assume $r \geq u(L)$, Then χ is universal over L since L is nonreal. Thus we find a vector $0 \neq \xi = (\xi_1, \ldots, \xi_r) \in L^r$ with $\chi(\xi) = 1 \in K$. Since $\operatorname{Im}\chi(\xi) = 0$ this implies that $\varphi \cong \operatorname{Im}\chi$ is isotropic over K: contradiction.

The special cases where $u(L) \leq 4$ follow from the main result together with 2.7.

As we will point out in the next section (Example 3.10) the estimates of Theorem 2.12 are sharp for $u(L) \leq 4$. But it is not known whether $u(K) = 6$ is possible if K is a field of transcendence degree 2 over a real-closed field. Several authors have worked on this problem but all attempts have failed. If in addition K is nonreal then $u(K) = 6$ would contradict Lang's Conjecture 2.4 from Chapter 5.

§3. Various Further Results

Our first concern is the behaviour of the u-invariant under a finite algebraic extension L/K. We start with estimates for

Going up

1. Let K be nonreal, let $[L : K] = d$. Then we have

3.1 Theorem. [Leep 1984].

$$u(L) \leq \frac{d + 1}{2}u(K).$$

This theorem will be proved in the next chapter by using systems of quadratic forms. (See Chapter 9, Corollary 2.2.)

2. Let now K be real. Then there is in general no upper estimate for the u-invariant of a finite extension L of K. This is shown by the following.

3.2 Examples.

a) Let $K = \mathbf{R}((x_1))\ldots((x_n))$, $L = K(\sqrt{-1})$. Then $p(K) = 1$, $u(K) = 0$, $u(L) = 2^n$ for all $n \in \mathbf{N}_0 \cup \infty$. In particular, it is possible that $[L : K] = 2$, $u(K) = 0$, $u(L) = \infty$.

b) For every $n \in \mathbf{N}$ there exists a real field K and a real quadratic extension $L = K(\sqrt{a})$ with

$$p(K) = 2, \ u(K) = 2^n, \ u(L) = \infty.$$

c) For every $n \in \mathbf{N}$ there exists a real field K such that $u(K) = 2^n$ and $u(L) = \infty$ for *all nonreal* fields L with $[L : K] < \infty$.

PROOF.

a) follows from Examples 1.2 and 1.7.

b) and c) are proved in the paper [Elman–Prestel 1984] by the method of intersecting several henselian closures. (Compare Proposition 1.5′ in Chapter 7.)

Next we consider estimates for u on

Going down

3.3 Theorem. Let $[L : K]$ be odd. Then

$$u(K) \leq u(L), \ u'(K) \leq u'(L).$$

PROOF. Both estimates are immediate from Springer's Theorem 1.12 in Chapter 6.

The next case which can be settled completely is the quadratic case $[L : K] = 2$. Following [Elman 1977] we work with u' instead of u. We get

3.4 Theorem. For an arbitrary field K (char $K \neq 2$) and every quadratic extension $L = K(\sqrt{a})$ of K we have

$$u'(K) \leq 3u'(L).$$

PROOF. (Sketch)

(0) We use some elementary but fundamental results about a quadratic extension $L = K(\sqrt{a})$:

a) q anisotropic quadratic form over K, $q \otimes L$ isotropic $\Longrightarrow$ $q \cong$ $\langle a_1, -aa_1, \dots \rangle$ for some $a_1 \in K^\bullet$.
q anisotropic over K, $q \otimes L \sim 0$ $\Longrightarrow$ $q \cong \langle 1, -a \rangle \otimes q_1$ for some q_1 over K.
In other words: The K-linear inclusion map $r : K \to L$ induces a $W(K)$-module homomorphism $r^* : W(K) \to W(L)$ with ker $r^* = \langle 1, -a \rangle W(K)$.

b) The K-linear map $s : L \to K$ given by $s(1) = 0$, $s(\sqrt{a}) = 1$ induces a $W(K)$-module homomorphism $s_* : W(L) \to W(K)$ – called trace or Scharlau transfer – with $\dim(s_*\varphi) = 2 \dim \varphi$.
$s_*\varphi$ isotropic over K $\iff$ φ represents an element of $K^\bullet$.
$s_*\varphi \sim 0$ over K $\iff$ $\varphi = r^*(q)$ for some form q over K for every quadratic form φ over L.

c) im $s_* =$ ker μ where μ is the multiplication with $\langle 1, -a \rangle$ in $W(K)$.

d) We have the exact triangle of Elman and Lam:

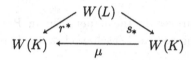

The proof of parts a) and b) is relatively easy, the nontrivial part of c) is proved by the same method as we used in the proofs of 2.9 and 2.12 for the special case $a = -1$ (where $s_*\varphi$ was denoted by Im φ); d) is an immediate consequence of a), b), c). For a full proof see [S, Ch. 2, Thm. 5.10].

(1) Let now q be an anisotropic form over K with $2 \times q \sim 0$. We will show by *induction* on $\dim q$

(1) $q \sim \langle 1, -a \rangle \otimes q_1 \oplus q_2$

where q_1, q_2 are quadratic forms over K such that $2 \times q_2 \sim 0$ and $q_2 \otimes L$ is anisotropic over L. By part (0) this is clear if $s(K) = 1$ (where the condition $2 \times q_2 \sim 0$ is automatically satisfied) or if $q \otimes L$ is anisotropic (take $q_1 = 0$) or if $q \otimes L$ is hyperbolic (take $q_2 = 0$).

Let us now asume $s(K) \neq 1$, $q \otimes L$ isotropic but not hyperbolic. From $2 \times q \sim 0$ we get: $\dim q$ is even. From (0) we get

$$q \cong a_1 \langle 1, -a \rangle \oplus q_*.$$

If $2 \times q_* \sim 0$ apply the induction hypothesis to q_*. If $2 \times q_* \not\sim 0$ then $q_* \cong q_3 \oplus q_4$ with $2 \times q_3 \sim 0$ and $\emptyset \neq 2 \times q_4$ anisotropic (see 2.3 and

2.4). Since q and q_3 are linear combinations of binary forms of the shape $\langle 1, -(1+v^2)\rangle$ we see that the discriminants $d = d(q)$ and $d_3 = d(q_3)$ are sums of two squares in K. The equation $q \cong a_1\langle 1, -a\rangle \oplus q_3 \oplus q_4$ then gives $d = ad_3 d(q_4)$, i.e. $ad(q_4) = t_4$ is also a sum of two squares. Furthermore we have

$$2 \times q \sim 2 \times a_1\langle 1, -a\rangle \oplus 2 \times q_4 \sim 0, \text{ i.e. } 2 \times q_4 \sim 2 \times (-a_1)\langle 1, -a\rangle.$$

Up to the scalar factor $-a_1$ the latter form is a 2-fold multiplicative form, hence either anisotropic or hyperbolic. Since $2 \times q_4$ is anisotropic and $\not\sim 0$ we must have

$$2 \times q_4 \cong 2 \times (-a_1)\langle 1, -a\rangle.$$

In particular we get $\dim q_4 = 2$, $q_4 \cong a_4\langle 1, -at_4\rangle$ where t_4 is a sum of two squares in K. This gives

$$q \cong a_1\langle 1, -a\rangle \oplus a_4\langle 1, -at_4\rangle \oplus q_3, \ 2 \times q_3 \sim 0.$$

The trick is now to use the relation

$$a_1\langle 1, -a\rangle \oplus a_4\langle 1, -at_4\rangle \sim a_1\langle 1, -a\rangle \otimes \langle 1, a_1 a_4 t_4\rangle \oplus a_4\langle 1, -t_4\rangle.$$

It implies

$$q \sim \langle 1, -a\rangle \otimes \langle a_1, a_4 t_4\rangle \oplus [a_4\langle 1, -t_4\rangle \oplus q_3].$$

Here the form $q_{**} := a_4\langle 1, -t_4\rangle \oplus q_3$ satisfies $2 \times q_{**} \sim 0$ and $\dim q_{**} = \dim q - 2$. Hence the inductive hypothesis can be applied to q_{**}. The proof of (1) is finished.

(2) The next useful observation is the equivalence $2 \times \langle 1, -a\rangle \cong \langle 1, -a\rangle \otimes \langle 1, -a\rangle$. Since $2 \times (\langle 1, -a\rangle \otimes q_1) \sim 2 \times (q \ominus q_2) \sim 0$ it implies

$$\langle 1, -a\rangle \otimes (\langle 1, -a\rangle \otimes q_1) \sim 0.$$

By part (0) this shows that $\langle 1, -a\rangle \otimes q_1 \sim s_* \varphi$ for some form φ over L. We get

(2) $\qquad q \sim s_* \varphi \oplus q_2$ with $2 \times q_2 \sim 0$, $\quad q_2 \otimes L$ anisotropic.

(3) The last step is to arrange $2 \times \varphi \sim 0$ in (2). Since $s_*(2 \times \varphi) \sim 2 \times s_* \varphi \sim 0$ we conclude from (0) that $2 \times \varphi$ represents an element $b \in K^{\bullet}$. Then $2 \times (\varphi \oplus \langle -b\rangle)$ is isotropic over L. Put $\psi := \varphi \oplus \langle -b\rangle$. Then $s_* \psi \sim s_* \varphi$ and $\psi \cong \omega \oplus \chi$ with $2 \times \omega \sim 0$ (over L) and $\dim \chi \leq \dim \psi - 2 < \dim \varphi$ by Proposition 2.3. We get $q \sim s_* \chi \oplus (s_* \omega \oplus q_2)$, $2 \times \omega \sim 0$, $2 \times q_2 \sim 0$. By induction on $\dim \varphi$ this shows that we can assume:

(3) $\qquad\qquad$ We have $2 \times \varphi \sim 0$ in (2).

The estimate for $u'(K)$ is now an immediate consequence of (3). Let q be anisotropic, $2 \times q \sim 0$. Let $q \sim s_* \varphi \oplus q_2$ as in (3). W.l.o.g. we may assume φ anisotropic over L. Then $\dim \varphi \leq u'(L)$ since $2 \times \varphi \sim 0$, and $\dim q_2 \leq u'(L)$ since $2 \times q_2 \sim 0$ and $q_2 \otimes L$ is anisotropic. This implies

$$\dim q \leq \dim(s_* \varphi) + \dim q_2 \leq 2u'(L) + u'(L) = 3u'(L).$$

Combining Theorems 3.1, 3.3 and 3.4 we get the following going-down result for the u-invariant.

3.5 Theorem. Let L/K be a finite field extension (char $K \neq 2$). Assume in addition that either L/K is normal or L is nonreal. Then $u(L) < \infty \implies u(K) < \infty$.

PROOF.

(1) If L/K is inseparable, let L_s be the separable closure of K in L. Then L/L_s is purely inseparable of p-power degree where p is an odd prime. For every $\alpha \in L$ we have $a = \alpha^{[L:L_s]} \in L_s$ and $\langle \alpha \rangle \cong \langle a \rangle$ over L. This shows that every quadratic form $\langle \alpha_1, \ldots, \alpha_n \rangle$ over L is isometric to a form $\langle a_1, \ldots, a_n \rangle$ coming from L_s and that $W(L_s)$ and $W(L)$ are naturally isomorphic. In particular $u(L) = u(L_s)$. If L/K is normal then L_s/K is a Galois extension, say of degree $2^r \cdot d'$ where d' is odd. By Galois theory we get a sequence of subfields

$$K \subset K_0 \subset K_1 \subset \ldots \subset K_r = L_s$$

such that $[K_0 : K] = d'$, $[K_\varrho : K_{\varrho-1}] = 2$ for $\varrho = 1, \ldots, r$. On applying 3.3 and 3.4 we get

$$u'(K) \leq u'(K_0) \leq 3^r \cdot u'(L_s) = 3^r u'(L) \leq 3^r u(L).$$

Since $u(K) \leq 2u'(K)$ the desired result follows.

(2) If L is nonreal, let M/K be the normal closure of L/K. By Theorem 3.1 we have $u(L) < \infty \implies u(M) < \infty$. Then part (1) applies to M/K and yields $u(K) < \infty$.

Since going up fails sometimes when starting from a real field L we are left with the following

3.6 Open problem. Let L/K be an arbitrary finite field extension (char $K \neq 2$). Is it always true that $u(L) < \infty$ implies $u(K) < \infty$?

More concrete finiteness results for the u-invariant $u(K)$ of a field K can be made if K satisfies some severe extra condition. By Chapter 2, statements

A and B at the end of §3, every quadratic form q over K gives rise to a Clifford algebra $C(q)$ and its class $c(q) \in B_2(K)$. For the multiplicative form $q = \langle 1, -a \rangle \otimes \langle 1, -b \rangle$ we have $c(q) = (a, b)$, a quaternion algebra. It is well-known that $c(q)$ is a (tensor) product of such quaternion algebras for every $q \in W(K)$. Let $\mathfrak{Q}$ be the *subset* of quaternion algebra classes (a, b) in $B_2(K)$. Then we have (without proof)

3.7 Theorem. Assume that $\mathfrak{Q}$ is closed under multiplication. Then

(1) $I^4(K)$ is torsion-free,

(2) $u(K) \in \{0, 1, 2, 4, 8\}$,

(3) $u(K) = 8 \iff I^3 \cap W_t \neq 0$.

The proof is elementary, but lengthy. See the original papers [Elman–Lam 1973_2, Thm. 3.4] and [Elman 1977, Thm. 4.7].

Note. From Merkurjev's theorem saying that $\bar{e}_2 : I^2 \to B_2(K)$ is surjective it follows that the above condition "$\mathfrak{Q}$ closed under multiplication" is in fact equivalent to the condition "$\mathfrak{Q} = B_2(K)$". But this is not needed for the proof of 3.7.

The most important progress in the investigation of the u-invariant after Leep's Theorem 3.1 is again due to A. Merkurjev (manuscripts from 1988/89, published paper [Merkurjev 1992]):

3.8 Theorem. (Merkurjev) Let $m \in \mathbb{N} \cup \infty$. Then there exist real and nonreal fields K with the properties:

$$u(K) = 2m, \quad I^3(K) \text{ torsion free}.$$

In particular, every even number is the u-invariant of a suitable field and the converse of 2.8 does not hold for $n \geq 3$.

Merkurjev's original proof makes use of deep results of Swan on the algebraic K-theory of quadrics. Further ingredients are a careful study of the *index-reduction* of division algebras under an extension $K \to K(q)$, where $K(q)$ is the *function field of the quadric* $q(x) = 0$, and finally a transfinite induction process. All the fields constructed are therefore highly infinite and complicated. Simpler proofs which avoid K-theory have been given by Lam [1989_1] for the case $u = 6$ and by Tignol [1990] for the general case.

Theorem 3.8 together with Proposition 1.3 and Note 1.6 leads us in a natural way to

3.9 Open Problem. Do there exist (nonreal) fields with *odd* u-invariant u, $1 < u < \infty$?

Following the method of Merkurjev other interesting results about the u-invariant have been obtained recently. I mention some of them without proof.

3.10 Examples. [Leep–Merkurjev 1994].

(1) For each even integer $n \geq 2$ there exists a quadratic extension L/K such that $u(K) = 2n$, $u(L) = 3n$.

(2) For each odd integer $n \geq 3$ there exists a quadratic extension L/K such that $u(K) = 2n$, $u(L) \in \{3n - 1, 3n\}$.

(3) For each positive integer n there exists a cyclic cubic extension L/K such that $u(K) = 2n$, $u(L) = 4n$.

In particular these examples show that the estimate $u(L) \leq \frac{d+1}{2} u(K)$ of Theorem 3.1 is sharp for $d = 1, 2, 3$. In contrast to this it is not known whether the estimate is sharp for $d > 3$. This is best expressed by

3.11 Open Problem. Let K be nonreal, let L/K be a finite field extension. Do we always have

$$u(L) \leq 2u(K)?$$

A considerable strengthening of 3.10 is provided by

3.12 Examples. [Mináč–Wadsworth 1993]. Let ℓ be an integer.

(1) If $\ell \geq 3$ there is a field K with $I^3(K) = 0$ and with quadratic extensions L_j for $\ell - 1 \leq j \leq \left[\frac{3\ell-1}{2}\right]$ such that $u(K) = 2\ell$ and $u(L_j) = 2j$.

(2) If $\ell \geq 2$ there is a field K with $I^3(K) = 0$ and with cyclic cubic extensions L_j for $\ell \leq j \leq 2\ell$ such that $u(K) = 2\ell$ and $u(L_j) = 2j$, $I^3(L_j) = 0$.

In particular (1) shows that u can "go down slightly" when going up to a quadratic extension L/K.

We finish our treatment of the u-invariant by a result and an open question about the u-invariant of a rational function field $K(t)$. We work with u' instead of u.

3.13 Proposition. Suppose char $K \neq 2$. Then

$$u'(K(t)) \geq 4 \sup_{[L:K]<\infty} \left[\frac{u'(L)}{2}\right].$$

PROOF. L/K be a finite field extension which we may assume to be separable. Then $L = K(\alpha)$ for a generating element $\alpha \in L$. Let $p(t)$ be the minimal polynomial of α, deg $p = [L : K] = d$. Every element $\beta \in L$ lifts to a unique polynomial $f(t) \in K[t]$ with deg $f < d$ such that $\beta = f(\alpha)$. Consider now any anisotropic quadratic form q_L over L of even dimension $2m$ with $2 \times q_L \sim 0$. By 2.9 we have

$$q_L \cong \beta_1 \langle 1, -(\gamma_1^2 + \delta_1^2) \rangle \oplus \ldots \oplus \beta_m \langle 1, -(\gamma_m^2 + \delta_m^2) \rangle.$$

By the definition of $u'(L)$ we can take $m = [\frac{u'(L)}{2}]$. Lift the elements β_i, γ_i, δ_i to polynomials f_i, g_i, h_i and consider the form

$$q = f_1 \langle 1, -(g_1^2 + h_1^2) \rangle \oplus \ldots \oplus f_m \langle 1, -(g_m^2 + h_m^2) \rangle$$

over $K(t)$. Clearly $2 \times q \sim 0$. Let v be the discrete valuation of $K(t)$ corresponding to p such that $v(p) = 1$. Then the form $q \oplus p(t)q$ over $K(t)$ has first and second residue forms (with respect to v) equal to q_L, and by Springer's theorem $q \oplus p(t)q$ is anisotropic over $K(t)$ (even over the completion of $K(t)$ with respect to v). (Compare the proof of 1.2 (4).) This gives $u'(K(t)) \geq 4m$.

3.14 Open Problem. Does $u(K) < \infty$ imply $u(K(t)) < \infty$ for all fields K?

This problem seems to be very difficult. Not even for a p-adic field $K = \mathbf{Q}_p$ or for a number field K is the answer known. Combining this question with Open Problem 3.11 and the examples of section 1 we should also mention

3.15 Open Problem. Let K/k be a finitely generated field extension where $k = \mathbf{Q}$ or $\mathbf{Q}_p$. Does this imply $u(K) < \infty$?

In the rest of this section we consider some other invariants which are related to the u-invariant and which have been introduced and studied in the hope of understanding the whole story better.

Next to u is the following invariant $\tilde{u}$:

3.16 Definition. $\tilde{u}(K) := \sup\{\dim \varphi : \varphi$ anisotropic and totally indefinite quadratic form over $K\}$ is called the *Hasse number* of K.
(Over a nonreal field K every form φ is said to be totally indefinite (since K has no orders). Over a real field K the empty form $\varphi = \emptyset$ is totally indefinite by definition. For $\varphi \neq \emptyset$ we must have $|\mathrm{sign}_P \varphi| < \dim \varphi$ for all orders P on K.)
It is clear that $\tilde{u}(K) = u(K)$ for all nonreal fields K and that $0 \leq \tilde{u}(K) \leq \infty$, $\tilde{u}(K) \neq 1$ and $u(K) \leq \tilde{u}(K)$ for a real field K. Example 1.4 shows that very often $\tilde{u}(K) = \infty$ for a real field K. Nevertheless it is of interest to investigate the conditions on K which give $\tilde{u}(K) < \infty$ and then to compare u and $\tilde{u}$. We may suppose K real. All results are given without proof.

3.17 Proposition. Under the assumption of 3.7 (Ω is a group) we have $\tilde{u}(K) < \infty$ and $\tilde{u}(K) = u(K) \leq 8$.

PROOF. See [Elman–Lam–Prestel 1973] and [Elman 1977].

3.18 Proposition.

$$\tilde{u}(K) < \infty \iff u(K) < \infty, \quad K \text{ is SAP}, \quad K \text{ has } (S_1).$$

PROOF. See [Elman–Prestel 1984].

The properties SAP (strong approximation property) and S_1 (stability) concern the order space $X_K = \{P : P \text{ order on } K\}$. They are very restrictive. I omit the precise definition of these properties.

3.19 Proposition. For each integer $m \geq 3$ there exists a real field K such that $I^3(K)$ is torsion-free and

$$\tilde{u}(K) = 2m, \quad u(K) = \tilde{u}(K) - 2.$$

This shows that u and $\tilde{u}$ can be both finite, but different. The proof uses Merkurjev's construction, see [Hornix 1991] or [Lam 1989₂].

Another invariant which is weaker but more easily accessible than the u-invariant is defined as follows.

3.20 Definition. $v(K) := \min\{n : I^n K \cap W_t(K) = 0\}$ is called the v-*invariant* of K.

From 2.8 we know that $u < 2^n$ (or already $u' < 2^n$) implies $v \leq n$. On the other hand 3.8 shows that $v < \infty$ (say $v = 3$) does not imply $u < \infty$. Let us state some interesting properties of v (without proof).

3.21 Theorem. Let K be nonreal, char $K \neq 2$, let L/K be a finite field extension, $[L : K] = d$.

a) $v(L) \leq v(K) + d - 1$.

b) If $d = 2$, then $v(L) = v(K)$.

c) If L/K is a Galois extension, then $v(K) \leq v(L)$.

d) $v(K) < \infty \iff v(L) < \infty$.

e) For suitable K and L we have $v(L) > v(K)$.

f) For $d \geq 2$ we have $v(L) \leq v(K) + 1 + \lfloor \log_2 \frac{d}{3} \rfloor$.

PROOF. a) – e) are proved in [Elman–Lam 1976]. The estimate f) which is considerably better than a) is proved in [Leep 1988].

3.22 Open Problem. Does $v(L) \leq v(K) + 1$ hold whenever $[L : K] < \infty$, K nonreal?

For real base field K there is in general no going-up result on the v-invariant, compare Example 3.2.

3.23 Proposition. Let K be a (real or nonreal) field of transcendence degree n over a real-closed field R. Then $v(K) \leq n + 1$.

PROOF. This is an immediate consequence of Example 1.2(6) and Theorem 2.12 together with the observation after Definition 3.20.

Finally Elman and Prestel [1984] introduced the following invariant $w(K)$:
3.24 Definition. $w(K) = u(K(\sqrt{-1}))$.

Clearly w is a weaker invariant than u since $u(K) < \infty$ does not imply $w(K) < \infty$ and since w eliminates the difficulties of reality. But w behaves much better under going up and going down. Clearly Theorems 3.3 and 3.5 imply

3.25 Proposition For every finite field extension L/K we have

$$w(L) < \infty \iff w(K) < \infty.$$

The invariant $w(K)$ will be used in the next chapter.

§4. Appendix: The case char $K = 2$

As we have already seen in Chapters 1 and 2 this case causes additional difficulties since we have to distinguish between symmetric bilinear forms resp. the Witt ring $W = W(K)$, and quadratic forms, resp. the W-module Wq. On the other hand many things simplify since K is nonreal with $-1 = 1^2$ and since every nontrivial element of W or Wq has additive order 2 (See Chapter 2, section 4). Concerning the u-invariant it is therefore clear that u', $\tilde{u}$ and w coincide with u whatever the definition will be. The basic definitions and results are due to [Baeza 1982].

4.1 Definition. For char $K = 2$ we put

(1) $u(K) := \sup\{\dim q : q \text{ anisotropic } regular \text{ quadratic form over } K\}$.

(2) $\hat{u}(K) := \sup\{\dim q : q \text{ anisotropic quadratic form over } K\}$.

(3) $v(K) := \min\{n : I^n Wq(K) = 0\}$ where I is the fundamental ideal of W.

It is immediately clear that we have $0 \leq u \leq \hat{u} \leq \infty$, u even (since every regular form has even dimension) and $1 \leq \hat{u}$ (since $\langle 1 \rangle$ is anisotropic). The invariant $\hat{u}$ is intimately related to the field degree $[K : K_0]$ where $K_0 = K^2$. Note also that, compared with Definition 3.20, there is a shift by 1 in the definition of the v-invariant, since the generators of $I^n Wq$ have dimension 2^{n+1}. This implies $2^{v(K)} \leq u(K)$ for char $K = 2$.

Let us now collect some important results about the invariants u, $\hat{u}$, v for fields of characteristic 2.

4.2 Proposition. $[K : K_0] \leq \hat{u}(K) \leq 2[K : K_0]$.

4.3 Proposition. $\hat{u} \neq 5$ and $\hat{u} \neq 2^n - 1$ for all $n \geq 2$.

4.4 Proposition. Let K/k be a *finitely generated* field extension of transcendence degree $n \geq 1$ (char $k = 2$).

a) If k is algebraically closed then

$$u(K) = \hat{u}(K) = [K : K_0] = 2^n, \quad v(K) = n.$$

b) If k is finite then

$$u(K) = \hat{u}(K) = 2[K : K_0] = 2^{n+1}, \quad v(K) = n + 1.$$

For the proof of 4.2 - 4.4 see [Mammone–Moresi–Wadsworth 1991]. The proofs of 4.2 and 4.3 are elementary, the proof of 4.4 is elementary modulo the use of the Tsen–Lang and Chevalley theorems (see Chapter 5).

4.5 Proposition. Let K be complete with respect to a discrete valuation with residue field k (char K = char $k = 2$). Then

$$u(K) = \hat{u}(K) = 2\hat{u}(k).$$

PROOF. See [Baeza 1982].

4.6 Proposition. Let t be transcendental over K. Then

$$2\hat{u}(K) \leq u(K(t)) \leq \hat{u}(K(t)) \leq 4\hat{u}(K).$$

PROOF. See [Mammone–Moresi–Wadsworth 1991].

4.7 Open Question. Does $u(K(t)) = \hat{u}(K(t))$ hold for all fields K of characteristic 2?

4.8 Proposition. Let L/K be a finite field extension. Then

a)

$$\frac{1}{2}\hat{u}(K) \leq \hat{u}(L) \leq 2\hat{u}(K).$$

b)
$$v(K) \leq v(L) \leq v(K) + 1 \quad \text{and} \quad v(L) = v(K) \text{ if } [L : K] = 2.$$

PROOF.

a) is proved in the paper of Mammone, Moresi and Wadsworth [1991].

b) is proved in [Aravire–Baeza 1989]. The proof is elementary but tricky. It is also shown that the inequality for $v(L)$ is best possible in general.

By means of valuation theory Mammone, Moresi and Wadsworth constructed fields K with $u(K) = 2^n$, $\hat{u}(K) = [K : K_0] = 2^m$ for every pair (n, m) of natural numbers with $n \leq m$. But this result is superseded by

4.9 Proposition. Let $n, m \in \mathbb{N}$. There exist fields K (char $K = 2$) with any of the following properties:

a) $u(K) = \hat{u}(K) = 6$.

b) $u(K) = 2n$, $\hat{u}(K) = \infty$ (for $n \geq 2$).

c) $u(K) = 2n$, $\hat{u}(K) = 2^m$ (for $m \geq n - 1$ and $2^m \geq 2n \geq 4$).

PROOF. See [Mammone–Tignol–Wadsworth 1991]. This proof uses Merkurjev's construction (see Theorem 3.8).

Summarizing these results we can say that Open Problems 3.6, 3.11, 3.14 and 3.22 are solved for char $K = 2$ if we replace u by $\hat{u}$. Open Problem 3.9 is partly solved for char $K = 2$, since we know from Proposition 4.3 that the infinitely many odd numbers of the form $2^n - 1$ cannot occur as $\hat{u}$-invariants. As at the end of Chapter 2 we realize that the theory for char $K = 2$ is ahead of the general theory!

Chapter 9

Systems of Quadratic Forms

§1. Some General Phenomena

Let K be a field, let r be a natural number. We consider r quadratic forms $q_1, \ldots, q_r$ over K in n common variables. This gives an (r-fold) *system* $q = (q_1, \ldots, q_r)$ over K. By the elementary considerations of Chapter 1 q may be identified with a *quadratic map* $q : V \to K^r$ where V is an n-dimensional K-vectorspace. Two systems $q = (q_1, \ldots, q_r)$, $q' = (q'_1, \ldots, q'_r)$ or their corresponding quadratic maps $q : V \to K^r$, $q' : V' \to K^r$ are called *equivalent*, $q \cong q'$, if there is a K-linear isomorphism $T : V \to V'$ such that $q'(Tv) = q(v)$ for all $v \in V$. The map $b : V \times V \to K^r$ with

$$b(v, w) = \frac{1}{2}(q(v + w) - q(v) - q(w))$$

is called the *symmetric bilinear map* corresponding to q. (If char $K = 2$ the factor $\frac{1}{2}$ has to be replaced by 1.) It is pretty clear that a given system q is not always equivalent to a system of diagonal forms. The *radical* rad $q = (V, q)^\perp$ is the subspace $\{w \in V : b(v, w) = 0 \text{ for all } v \in V\}$. Obviously we have

$$\text{rad } q = \text{rad } q_1 \cap \ldots \cap \text{rad } q_r.$$

q is called *regular* if rad $q = 0$.

$$O(q) = \{T \in GL(V) : q(Tv) = q(v) \text{ for all } v \in V\}$$

is called the *orthogonal group* of the quadratic map q.

If $q^{(\mu)} : V^{(\mu)} \to K^r$ ($\mu = 1, \ldots, m$; $\dim V^{(\mu)} = n_\mu$) are several quadratic maps (in disjoint sets of variables) we can form the *direct orthogonal sum*

$$q = q^{(1)} \oplus \ldots \oplus q^{(m)} : V^{(1)} \oplus \ldots \oplus V^{(m)} \to K^r.$$

It has $n = n_1 + \ldots + n_m$ variables. Clearly q is regular if and only if all $q^{(\mu)}$ are regular. In particular every $q : V \to K^r$ induces its *multiples*

$$m \times q = \underbrace{q \oplus \ldots \oplus q}_{m} : m \times V = \underbrace{V \oplus \ldots \oplus V}_{m} \to K^r.$$

1.1 Definition. Let $q : V \to K^r$ be a quadratic map.

a) q is called *isotropic* if $q(v) = 0$ for some $0 \neq v \in V$.

b) q is called *metabolic*, $q \sim 0$, if there exists a subspace $U \subset V$ with $\dim U \geq \frac{1}{2} \dim V$ such that $q|U = 0$.

1.2 Notes.

(1) The definition of metabolic maps and the sign $\sim$ are crucial. If char $K \neq 2$ we could restrict ourselves to considering only regular maps q. Then $q \sim 0$ implies $\dim V$ even and $\dim U = \frac{1}{2} \dim V$. But it does not follow that b) implies one of the following conditions:

b') $V = U_1 + U_2$ with subspaces U_1, U_2 such that $q|U_1 = q|U_2 = 0$.

b'') All maximal totally isotropic subspaces of V have the same dimension, namely $\frac{1}{2} \dim V$.

b''') For every two maximal totally isotropic subspaces U_1, U_2 of V there is an element $T \in O(q)$ with $U_2 = T(U_1)$.

b) is the weakest possible condition for $q \sim 0$. In agreement with the terminology in Chapter 2, section 4 for one quadratic form in the case char $K = 2$ we call such a map q metabolic instead of hyperbolic. It is not yet clear which one of the possible definitions for the sign $\sim$ is best suited to the treatment of systems of quadratic forms. But fortunately it turns out that every anisotropic system q with $2 \times q \sim 0$ satisfies the seemingly stronger property $q \cong -q$ (see Proposition 1.13 below).

(2) In contrast to the theory of one quadratic form we have no Witt group of quadratic maps. The relation $m \times q \sim 0$ does not mean that "q has order m" in some abelian group. It could well happen that $2 \times q \not\sim 0$, but $3 \times q \sim 0$ and $4 \times q \sim 0$.

(3) A system $q = (q_1, \ldots, q_r)$ of quadratic forms induces a *pencil* of quadratic forms, namely the set $\{\lambda_1 q_1 + \ldots + \lambda_r q_r : \lambda_\varrho \in K\}$. Many properties which we investigate for systems in this chapter have analogues for pencils. But the details are slightly different as can already be seen from the definition of equivalence for pencils which allows a further linear transformation in the image space K^r. In order not to overload the presentation I leave out pencils completely.

As in the case $r = 1$ we have

1.3 Lemma. If char $K \neq 2$ and $q = (q_1, \ldots, q_r)$ is an anisotropic system over K then q is regular.

PROOF. $0 \neq v \in \text{rad } q$ would imply $b(v, w) = 0$ for all $w \in V$, hence $q(v) = b(v, v) = 0$: contradiction.

1.4 Lemma. If K has finite level $s = s(K)$ then $2s \times q \sim 0$ for every system q over K.

PROOF. Let $q : V \to K^r$ be the given quadratic map.

a) If $s = 1$ then $-1 = a^2$ with $a \in K$. Consider the subspace $U = \{(v, av) \in V \oplus V : v \in V\}$ of $V \oplus V$. We have $\dim U = \dim V = \frac{1}{2} \dim(V \oplus V)$ and $(2 \times q)(u) = q(v) + q(av) = q(v) + a^2 q(v) = 0$ for each $u = (v, av) \in U$. Therefore $2 \times q \sim 0$.

b) If $s > 1$ then in particular char $K \neq 2$. Let $q_0 = 2s \times \langle 1 \rangle \cong s \times H$ be the hyperbolic quadratic form on $V_0 = K^{2s}$. It vanishes identically on a subspace $U_0 \subset V_0$ with $\dim U_0 = s$. Let $q = (q_1, \ldots, q_r) : V \to K^r$ be the given system. Then $2s \times q$ can be identified with $q_0 \otimes q = (q_0 \otimes q_1, \ldots, q_0 \otimes q_r) : V_0 \otimes V \to K^r$. From this it is immediate that $q_0 \otimes q$ vanishes identically on the subspace $U_0 \otimes V$ of $V_0 \otimes V$. Hence $2s \times q \sim 0$.

Our main interest in systems of quadratic forms comes from their intimate connection to questions about the u-invariant. It is therefore very natural that we define "system-u-invariants" u_r and u'_r as follows:

1.5 Definition. For a field K and a natural number r we put
$u_r := u_r(K) := \sup\{n : \text{ there exists an anisotropic quadratic map } q : K^n \to K^r \text{ with } m \times q \sim 0 \text{ for some } m \in \mathbb{N}\}$.
$u'_r := u'_r(K) := \sup\{n : \text{ there exists an anisotropic quadratic map } q : K^n \to K^r \text{ with } 2 \times q \sim 0\}$.

If K is nonreal we will work with u_r but if K is real we will work with the modified invariant u'_r as we know from the last chapter that this simplifies things quite a bit. Unfortunately, however, I have not been able to prove that $u'_r < \infty$ implies $u_r < \infty$ for $r > 1$ if K is real.

Of course we are interested in upper estimates for the invariants u_r, u'_r. But to begin with let us state the following trivial result.

1.6 Lemma.

(1) $u_1(K) = u(K)$, $u'_1(K) = u'(K)$ for char $K \neq 2$.

(2) $u_1(K) = u'_1(K) = \hat{u}(K)$ for char $K = 2$.

(3) $u_r(K) \geq r u_1(K)$, $u'_r(K) \geq r u'_1(K)$ for all $r \in \mathbb{N}$ and fields K.

PROOF.

(1) and (2) follow from Definitions 1.5, 2.1, 4.1 in Chapter 8.

(3) Let q_1 be an anisotropic quadratic form over K in $n = u_1(K)$ (resp. $n = u'_1(K)$) variables which satisfies $m \times q_1 \sim 0$ for some $m \in \mathbb{N}$ (for $m = 2$). Iterate q_1 r times with different sets of variables. This leads to an r-fold system q in $r \cdot n$ variables which is obviously anisotropic and satisfies $m \times q \sim 0$ (resp. $2 \times q \sim 0$).

Concerning the behaviour of u_r, u'_r under finite extensions we have

1.7 Proposition. Let L/K be a finite field extension of degree d. Let $q : V \to L^r$ be a quadratic map where $\dim_L V = n$. By reducing constants from L to K q induces a quadratic map $q_{(K)} : V_{(K)} \to K^{rd}$. We have:

(1) q isotropic over L $\iff$ $q_{(K)}$ isotropic over K.

(2) $m \times q \sim 0$ over L $\iff$ $m \times q_{(K)} \sim 0$ over K (for each $m \in \mathbb{N}$).

(3) $u_r(L) \leq \frac{1}{d} u_{rd}(K)$, $u'_r(L) \leq \frac{1}{d} u'_{rd}(K)$.

PROOF.

(1) $V_{(K)}$ is the set V considered as a K-vectorspace of dimension $\dim_K V_{(K)} = d \dim_L V = n \cdot d$, $q_{(K)}(v) := q(v)$ for $v \in V_{(K)} = V$ may be considered as a system of rd quadratic forms over K by identifying L^r with K^{rd}. Then statement (1) is clear. (For a different and more explicit proof compare Theorem 1.3 of Chapter 5.)

(2) Put $\hat{q} = m \times q$, $\hat{V} = m \times V$, let $\hat{U}$ be an L-subspace of $\hat{V}$. If $\hat{q}|\hat{U} = 0$ and $\dim_L \hat{U} \geq \frac{1}{2} \dim \hat{V} = \frac{1}{2} mn$ then clearly $\hat{q}_{(K)}|\hat{U}_{(K)} = 0$ where $\dim_K \hat{U}_{(K)} = d \cdot \dim_L \hat{U} \geq \frac{1}{2} mnd = \frac{1}{2} \dim \hat{V}_{(K)}$. This shows $\hat{q}_{(K)} = m \times q_{(K)} \sim 0$. Conversely, assume that $\hat{q}_{(K)}$ vanishes on a K-subspace $U \subset m \times V_{(K)} = \hat{V}_{(K)}$ with $\dim U \geq \frac{1}{2} mnd$. Since $\hat{q}$ and its associated symmetric bilinear map $\hat{b}$ are quadratic and bilinear respectively with respect to L it follows that $\hat{q} = \hat{q}_{(K)}$ actually vanishes on the L-subspace $\tilde{U} \subset m \times V$ which is generated by U. Then $\dim_L \tilde{U} = \frac{1}{d} \dim_K \tilde{U} \geq \frac{1}{d} \dim_K U \geq \frac{1}{2} mn$. Therefore $\hat{q} = m \times q \sim 0$.

(3) follows immediately from (1) and (2).

1.8 Notes.

(1) It is desirable to prove similar going-up results for the cases $L = K(t)$, $L = K((t))$ or L a field with a complete discrete valuation v and residue field K. But only very little is known up to now.

(2) Essentially the only "good" case is the following:

(∗) K is nonreal and $u_r(K)$ is bounded by a linear function of r, say $u_r(K) \leq c(K) \cdot r$ where $c(K)$ is some real constant.

Then the methods of Chapter 5 (restricted to systems of quadratic forms) apply: Choose an integer i such that $c(K) \leq 2^i$. Then K is a *quadratic C_i-field*, i.e. every system of r quadratic forms over K in $n > r \cdot 2^i$ variables is isotropic (compare Definition 2.8 of Chapter 5). The proofs of Theorems 1.4 and 2.2 of Chapter 5 may then be specialized to forms of degree 2 and yield: $K(t)$ and $K((t))$ are quadratic C_{i+1}-fields, in particular $u_r \leq 2^{i+1} \cdot r$ for these two fields (and each $r \in \mathbb{N}$).

(3) The case where L has residue field K is already much more difficult even if L is supposed to be nonreal. This is best illustrated by the Restricted Artin Conjecture 2.5(d) of Chapter 5. At least we know that $u_r(\mathbb{Q}_p)$ is finite for all $r \in \mathbb{N}$ and all primes p. In the next section (Note 2.3(3)) we shall prove

$$u_r(\mathbb{Q}_p) \leq 2r(r+1).$$

But this estimate is not linear in r.

(4) If, however, K and L are real or if $u_r(K)$ is not bounded by a linear function of r then nearly nothing is known about the going-up problem for the invariants u_r and u'_r. The substitution method for the proof of Theorems 1.4 and 2.2 of Chapter 5 does not work because we do not know whether the condition $m \times q \sim 0$ for the given system q over L implies a similar condition for the (various) induced systems over K which occur during the "intended proof". And if $u_r(K)$ is not bounded by a linear function of r the methods of Chapter 5 fail completely even if K is nonreal. Compare Open Problem 3.14 in Chapter 8.

(5) The "easiest" unsolved going-up problem for the u_r-invariants might be the following: Suppose K nonreal, $L = K((t))$. Is it true that

$$u_r(L) \leq 2u_r(K) \quad \text{for all } r \in \mathbb{N}?$$

Another result from the theory of quadratic forms which does not carry over to systems of quadratic forms is Springer's theorem on odd degree extensions (see Chapter 6, Theorem 1.12). This will be shown by the following.

1.9 Examples. Let $K = \mathbb{Q}$, $L = \mathbb{Q}(\sqrt[3]{2})$. The following systems of quadratic forms are anisotropic over K but isotropic over L:

(1) $q_1 = x_1^2 - 2x_2x_3, \quad q_2 = x_2^2 - x_1x_3, \quad q_3 = 2x_3^2 - x_1x_2 \quad (n = r = 3)$.

(2) Take ten variables x_j, y_j $(1 \leq j \leq 4)$, s, t and the seven quadratic forms

$$\begin{aligned}
q_j &= y_j t - x_j^2 \quad (1 \leq j \leq 4), \\
q_5 &= st - x_1 x_2, \\
q_6 &= x_1 y_1 - 2x_2 y_2 + 7(x_3 y_3 - 2x_4 y_4), \\
q_7 &= t^2 - 6s^2 - 3y_1^2 - 9y_2^2 - y_3^2 - y_4^2.
\end{aligned}$$

PROOF.

(1) If $(x_1, x_2, x_3) \neq (0,0,0)$ and $q_1(x) = q_2(x) = q_3(x) = 0$ then $x_1 x_2 x_3 \neq 0$ and $x_1^3 = 2x_1 x_2 x_3 = 4x_3^3$, $x_2^3 = x_1 x_2 x_3 = 2x_3^3$, hence $(x_1, x_2, x_3) = x_3(\sqrt[3]{4}, \sqrt[3]{2}, 1)$. Thus we have a nontrivial solution in L but none in K.

(2) It is easily seen that the system $q_1 = \ldots = q_7 = 0$ has no nontrivial solution with $t = 0$ (in a real field like K or L). If, say, $t = 1$ then $y_j = x_j^2$ and $q_6 = 0$ implies $x_1^3 - 2x_2^3 + 7(x_3^3 - 2x_4^3) = 0$. This equation has no integral solution mod 7 except the trivial one since $x^3 \equiv 0, \pm 1$ mod 7 for all $x \in \mathbf{Z}$. Over $L = \mathbf{Q}(\sqrt[3]{2})$ a nontrivial solution of our system can be constructed by taking $x_1 = \delta$, $x_2 = 1$, $x_3 = x_4 = 0$, $t = 1 + \delta$ where $\delta := \sqrt[3]{2}$. Then $(1 + \delta)y_1 = \delta^2$, $(1 + \delta)y_2 = 1$, $y_3 = y_4 = 0$ and $(1 + \delta)s = \delta$ follow from $q_1 = \ldots = q_5 = 0$. $q_6 = 0$ is clearly satisfied and $q_7 = 0$ follows on noting that $(1 + \delta)^4 = 6\delta^2 + 3\delta^4 + 9$.

Example (1) is simple but not very satisfying since from a geometric point of view one should always assume $r < n$. Example (2) which is from [Cassels 1979] meets this requirement. By further extending example (2) Cassels constructed examples which have infinitely many solutions over L or where the difference $n - r$ is arbitrarily large. See his original paper. A more geometric and systematic investigation of the intersection of three quadrics is due to [Coray 1980].

In view of Examples 1.9 one might ask why we don't take $r = 2$ and a suitable n. But it turns out that this is not possible because we have

1.10 Proposition. (Amer 1976, Brumer 1978, independently) Let K be a field, let t be an indeterminate over K, let (q_1, q_2) be a pair of quadratic forms over K in n common variables, say $q_\varrho(x) = \sum\limits_{i,j=1,\ldots,n} a_{ij}^{(\varrho)} x_i x_j$ $(\varrho = 1, 2)$.

Put $q = q_1 + tq_2$, i.e. $q(x) = \sum\limits_{i,j=1,\ldots,n} (a_{ij}^{(1)} + t a_{ij}^{(2)}) x_i x_j$. This is an n-dimensional quadratic form over $K(t)$ whose coefficients are polynomials of degree ≤ 1. q is isotropic over $K(t)$ if and only if (q_1, q_2) is isotropic over K.

PROOF. If the pair (q_1, q_2) has a common nontrivial solution $0 \neq u \in K^n$ then clearly $q(u) = 0$. Conversely assume that we have a vector $0 \neq u \in K(t)^n$ with $q(u) = 0$. W.l.o.g. we can assume that $u = u_0 + tu_1 + \ldots + t^d u_d$ is a polynomial vector with $u_i \in K^n$, $u_0 \neq 0$, $u_d \neq 0$. If $q(u_0) = 0$ then we immediately get $q_1(u_0) = q_2(u_0) = 0$, i.e. the pair (q_1, q_2) is isotropic over K. If $q(u_0) \neq 0$ then $d \geq 1$. Here we shall produce another polynomial vector $0 \neq u''$ of degree $\leq d - 1$ with $q(u'') = 0$. By induction on d this will finish the proof. (It is no surprise that this proof is very similar to the proof of 2.2 in Chapter 1.)

Let $b(v, w) := q(v + w) - q(v) - q(w)$ be the bilinear form corresponding to q. (We choose this definition in order to avoid any distinction as regards the characteristic of K.) By elementary calculation we find for $c \in K(t)$ and $v, w \in K(t)^n$:

(1) If $q(w) \neq 0$ then $q(v + cw) = q\left(v - (c + \frac{b(v,w)}{q(w)})w\right)$.

(2) If in addition $c \neq 0$ and $q(v + cw) = 0$ then $c + \frac{b(v,w)}{q(w)} = -\frac{1}{c}\frac{q(v)}{q(w)}$ and $q(v + \frac{1}{c}\frac{q(v)}{q(w)}w) = 0$.

We now start with the given equation $q(u) = 0$ and put $u' = u_1 + tu_2 + \ldots + t^{d-1}u_d$. Then $u = u_0 + tu'$ and $q(u' + \frac{1}{t}u_0) = 0$ in $K(t)$. Put $v = u'$, $w = u_0$, $c = \frac{1}{t}$ in (1) and (2). We get $q(u'') = 0$ where $u'' := u' + t\frac{q(u')}{q(u_0)}u_0$. It remains to show that the scalar $t\frac{q(u')}{q(u_0)} \in K(t)$ is a polynomial of degree $\leq d - 1$.

By specializing $t \to 0$ in the equation $q(u) = 0$ we see that $q_1(u_0) = 0$, $q_2(u_0) \neq 0$ and $q(u_0) = tq_2(u_0)$. In other words: $\frac{t}{q(u_0)} = \frac{1}{q_2(u_0)} \in K$. Furthermore $q(u') = q(\frac{u-u_0}{t}) = \frac{1}{t^2}(-b(u, u_0) + q(u_0))$ is a polynomial of degree at most $d + 1 - 2 = d - 1$ since b is linear in u and the coefficients of b are of degree ≤ 1.

1.11 Corollary. Let L/K be a finite field extension of odd degree. If a pair (q_1, q_2) of quadratic forms over K has a nontrivial common zero in L then it has such a zero in K.

PROOF. Apply Springer's theorem to the form $q_1 + tq_2$ over $K(t)$, then apply Proposition 1.10.

1.12 Note. Proposition 1.10 would not hold if we replaced the rational function field $K(t)$ by the power series field $K((t))$: Take $q_1 = x_1^2 - x_2^2$, $q_2 = -x_2^2$. The system (q_1, q_2) is clearly anisotropic over K but the form $q = q_1 + tq_2 = x_1^2 - (1 + t)x_2^2$ is isotropic over $K((t))$ since $1 + t$ is a square in $K((t))$ provided char $K \neq 2$.

We finish this introductory section on systems with the proof of the statement which was mentioned at the end of Note 1.2(1). This result will be needed in section 3.

1.13 Proposition. Let K be a field, and also V be an n-dimensional K-vectorspace, and let $q = (q_1, \ldots, q_r) : V \to K^r$ be an *anisotropic* quadratic map with $2 \times q \sim 0$. Then there exists $T \in GL(V)$ such that $q(Tv) = -q(v)$ for all $v \in V$.

PROOF. By assumption we have $V \oplus V = U \oplus W$ with $\dim U = \dim W = n$ and $2 \times q|_U = 0$. Let $u = v_1 \oplus v_2$ be the unique decomposition of an element $u \in U$ ($v_1, v_2 \in V$). The maps $T_i : U \to V$ with $T_i(u) = v_i$ ($i = 1, 2$) are K-linear. They are also injective: If $v_2 = 0$ then $(2 \times q)(u) = q(v_1) + q(v_2) = q(v_1) = 0$, hence $v_1 = 0$ since q is anisotropic, hence $u = 0$. Therefore T_1, T_2 are bijective. Put $T := T_2 \circ T_1^{-1} \in GL(V)$. Then $Tv_1 = v_2$ and $q(v_1) + q(Tv_1) = q(v_1) + q(v_2) = (2 \times q)(u) = 0$. Since T_1 is surjective this holds for every $v_1 \in V$. This finishes the proof. If char $K \neq 2$ then $U = \{(v, Tv) : v \in V\}$ and $U' = \{(v, -Tv) : v \in V\}$ are two totally isotropic subspaces of $(2 \times V, 2 \times q)$ with $\dim U = \dim U' = n$ and $U \cap U' = 0$, i.e. $2 \times q$ is "hyperbolic".

§2. Leep's Theorem

Despite difficulties in proving "going-up" results for systems of r quadratic forms it turns out that over a nonreal field K it is possible to fix K and use induction on r. Thereby we obtain a quite satisfactory upper estimate for u_r depending on u_1 and r. Combining this with Proposition 1.7 we get a proof of Theorem 3.1 of the last chapter which is the best possible going-up result for the u-invariant under a finite extension L/K known today.

2.1 Theorem. [Leep 1984] Let K be a nonreal field with finite invariant $u_1 = u_1(K)$. Then the system-u-invariant $u_r = u_r(K)$ satisfies the estimates

(1)
$$u_r \leq ru_1 + u_{r-1} \text{ for } r \geq 2,$$

(2)
$$u_r \leq \frac{r(r+1)}{2} u_1 \text{ for all } r \in \mathbb{N}.$$

PROOF.

(1) Let $q = (q_1, \ldots, q_r) : V \to K^r$ be anisotropic where V is a K-vectorspace of dimension n. Define $b : V \times V \to K^r$ by $b(v_1, v_2) = q(v_1 + v_2) - q(v_1) - q(v_2)$. Let U be a *subspace* (of V) of *maximal* dimension such that $(q_2, \ldots, q_r)|U \equiv 0$. Then b_ϱ vanishes on $U \times U$ for $\varrho = 2, \ldots, r$. Since $q|U$ must be anisotropic it is clear that $q_1|U$ is anisotropic. Hence $\dim U \leq u_1$. Let now $W_\varrho = \{v \in V : b_\varrho(v, U) = 0\}$ ($\varrho = 2, \ldots, r$) be the orthogonal space of $U \subseteq V$ with respect to the bilinear form b_ϱ. The condition $b_\varrho(v, U) = 0$ for $v \in W_\varrho$ amounts to a system of $\dim U$

K-linear equations for $v \in V$. This shows that $\dim W_\varrho \geq n - \dim U$. Put $W := W_2 \cap \ldots \cap W_r$. Then $\dim W \geq n - (r - 1)\dim U$. Furthermore $U \subseteq W$. Let W_0 be any complement of U in W, such that $W = U \oplus W_0$. We claim that the system $(q_2, \ldots, q_r)|W_0$ must be anisotropic. Otherwise there would be a vector $0 \neq w_0 \in W_0$ with $q_2(w_0) = \ldots = q_r(w_0) = 0$. Put $U' = U \oplus K w_0$ and let $v = u + c w_0$ be an arbitrary vector in U' $(u \in U, c \in K)$. Then $q_\varrho(v) = q_\varrho(u) + c^2 q_\varrho(w_0) + c b_\varrho(u, w_0) = 0 + 0 + 0 = 0$ for $\varrho = 2, \ldots, r$, i.e. $(q_2, \ldots, q_r)|U' \equiv 0$. This contradicts the maximality of U. Hence $(q_2, \ldots, q_r)|W_0$ is anisotropic and $\dim W_0 \leq u_{r-1}$. Combining this with the estimate $\dim W_0 = \dim W - \dim U \geq n - r \dim U$ we get

$$n = \dim V \leq r \dim U + \dim W_0 \leq r u_1 + u_{r-1}.$$

This holds for any anisotropic pair (V, q), hence (1) follows from the definition of u_r.

(2) follows from (1) by trivial induction on r.

2.2 Corollary. Let K be nonreal, let L/K be a finite field extension of degree d. Then

$$u_1(L) \leq \frac{d+1}{2} u_1(K),$$

where $u_1 = u$ for char $K \neq 2$, $u_1 = \hat{u}$ for char $K = 2$ by Lemma 1.6.

PROOF. By Proposition 1.7(3) we have $u_1(L) \leq \frac{1}{d} u_d(K)$. Together with 2.1(2) for $r = d$ this gives

$$u_1(L) \leq \frac{d+1}{2} u_1(K).$$

2.3 Notes.

(1) The estimates of Theorem 2.1 and Corollary 2.2 are best possible for $r \leq 3$ and $d \leq 3$: For $r = 2$ Example 2.7 of Chapter 5 shows that the system $x^2 - yz$, $y^2 + xz + z^2$ is anisotropic over the quadratic closure K of $\mathbf{Q}$. Hence $u_1(K) = 1$, $u_2(K) = 3 = \frac{2 \cdot 3}{2} u_1(K)$. It is known that K has a cubic extension L with infinite square class group $L^\bullet / L^{\bullet 2}$ (see [L, App. to Ch. 7, Cor. 3]). This implies $u_1(L) > 1$, therefore $u_1(L) = 2$ by Corollary 2.2. Hence estimate 2.2 is sharp for $d = 3$ and a priori estimate 2.1 (2) must be sharp for $r = 3$ and the field K. The proof for $d = 2$ is more difficult. It is covered by Example 3.10 (1) of Chapter 8.

(2) It is not known whether the estimates are sharp for $r \geq 4$ and $d \geq 4$. As already noticed in Note 1.8 (2) it would be of fundamental importance to know whether or not a linear estimate like

$$u_r(K) \leq r \cdot c(K) u_1(K)$$

holds for every nonreal field K and a suitable constant $c(K)$.

(3) Even for the classical fields $K = \mathbf{Q}_p$ or K a nonreal number field the estimates for $u_r(K)$ following from Theorem 2.1 and $u_1(K) = 4$, $u_2(K) = 8$ (see 2.5(d) and 2.6 in Chapter 5) are much better than all previously known estimates.

Theorem 2.1 can be generalized without difficulties to λ-hermitian forms over a skew-field. We consider the following situation: L is a *skew-field* with centre Z and with *involution* $*$ (i.e. $(a + b)^* = a^* + b^*$, $(ab)^* = b^*a^*$ and $(a^*)^* = a$ for all $a, b \in L$). λ is a fixed element with $\lambda \in Z$ and $\lambda\lambda^* = 1$. V is a finite-dimensional left L-vectorspace. h is a λ-*hermitian form*, i.e. the map $h : V \times V \to L$ has the two properties

(i) $h(x, y)$ is L-linear w.r.t. the first component x, for every $y \in V$.

(ii) $h(x, y) = \lambda h(y, x)^*$ for all $x, y \in V$.

(For the general theory of λ-hermitian forms see [S, Ch. 7, §6].) h is called *anisotropic* if $x \in V$, $h(x, x) = 0$ imply $x = 0$. Similarly, a *system* $h = (h_1, \ldots, h_r) : V \times V \to L^r$ of r λ-hermitian forms on V is called anisotropic if $x \in V$, $h(x, x) = 0$ imply $x = 0$.

2.4 Definition. $u_r(L, *, \lambda) := \sup\{\dim V :$ there exists an anisotropic λ-hermitian map $h : V \times V \to L^r\}$.

Of course the new invariant $u_r(L, *, \lambda)$ will in general depend on all the parameters $r, L, *, \lambda$, and is not necessarily finite. For instance, let $L = \mathbf{C}$, $* = $ complex conjugation, $\lambda = 1$, $r = 1$. Take $V = L^n$ and $h(x, y) := x\bar{y} := \sum_{i=1}^{n} x_i\bar{y}_i$. Then $h(x, x) = \sum_{1}^{n} |x_i|^2 = 0$ implies $x = 0$. Hence h is anisotropic for all n.

2.5 Theorem. Assume that $u_1(L, *, \lambda)$ is finite. Then

$$u_r(L, *, \lambda) \leq ru_1(L, *, \lambda) + u_{r-1}(L, *, \lambda)$$

for all $r \geq 2$, hence $u_r(L, *, \lambda) \leq \frac{r(r+1)}{2}u_1(L, *, \lambda)$ for all $r \in \mathbf{N}$.

PROOF. We imitate the proof of 2.1: Let $h : V \times V \to L^r$ be anisotropic. For *any* subspace U of V the restriction of h to $U \times U$ is a λ-hermitian map on $U \times U$. Let U be maximal such that $(h_2, \ldots, h_r)|U \times U \equiv 0$. Then h_1 is anisotropic on $U \times U$, hence $\dim U \leq u_1(L, *, \lambda)$. For $\varrho \in \{2, \ldots, r\}$ the set $W_\varrho = \{v \in V : h_\varrho(v, U) = 0\}$ is a left L-subspace of V since h_ϱ is L-linear in the first argument. Furthermore it is enough to check the equation $h_\varrho(v, U) = 0$ on a basis of U, hence $\dim W_\varrho \geq \dim V - \dim U$. Put $W := W_2 \cap \ldots \cap W_r$. Then $\dim W \geq \dim V - (r-1)\dim U$, and $U \subseteq W$ since $U \subseteq W_\varrho$ for $\varrho = 2, \ldots, r$. Let W_0 be any complement of U in W. Then $\dim W_0 \geq \dim V - r\dim U$. We claim that $(h_2, \ldots, h_r)$ must be anisotropic on W_0. Otherwise we have

$0 \neq w_0 \in W_0$ with $h_\varrho(w_0, w_0) = 0$ for $\varrho = 2, \ldots, r$. Consider any pair $x_1, x_2 \in U' := U \oplus Lw_0$, say $x_i = u_i + a_i w_0$ with $u_i \in U$, $a_i \in L$ $(i = 1, 2)$. Then we have for $\varrho \in \{2, \ldots, r\}$:

$$
\begin{aligned}
h_\varrho(x_1, x_2) &= h_\varrho(u_1 + a_1 w_0, u_2) + h_\varrho(u_1, a_2 w_0) + h_\varrho(a_1 w_0, a_2 w_0) \\
&= h_\varrho(u_1 + a_1 w_0, u_2) + \lambda h_\varrho(a_2 w_0, u_1)^* + a_1 h_\varrho(w_0, w_0) a_2^* \\
&= 0 + 0 + 0 = 0
\end{aligned}
$$

since $u_1 + a_1 w_0 \in W$, $a_2 w_0 \in W$ and $h_\varrho(w_0, w_0) = 0$. (Be careful with the order of factors in the skew-field L !). The rest of the proof is as before: We have just shown that the existence of w_0 implies $(h_2, \ldots, h_r)|U' \times U' \equiv 0$ which is in contradiction to the maximality of U. Therefore $(h_2, \ldots, h_r)$ is anisotropic on W_0, and $\dim W_0 \leq u_{r-1}(L, *, \lambda)$. This gives $\dim V \leq r \dim U + \dim W_0 \leq r u_1(L, *, \lambda) + u_{r-1}(L, *, \lambda)$ and the desired estimate for $u_r(L, *, \lambda)$.

Note that 2.1 follows from 2.5 for $L = K$, $* = \mathrm{id}$, $\lambda = 1$ in the case char $K \neq 2$ where $b(x, x) = 0$ is equivalent to $q(x) = 0$, but not in the case char $K = 2$. We shall need a special case of Theorem 2.5 in the next section.

§3. Systems over Real Fields

If we want to prove a result like Leep's theorem for the invariants u_r or u'_r of a *real* field F we are faced with a serious problem: It is necessary to restrict a given system $q = (q_1, \ldots, q_r)$ on V to various subspaces $U, W, \ldots$ of V. But a property like $2 \times q \sim 0$ is not invariant under a restriction map $q \to q|U$. Thus the main idea of Leep's induction proof breaks down. In the paper [Pfister 1989] I have shown how it is possible to circumvent this difficulty for systems q with $2 \times q \sim 0$. These systems q are related to certain systems h of λ-hermitian forms over suitable extension fields L of F. Then Theorem 2.5 can be applied to h and the result can be translated back into a finiteness result for $u'_r(F)$ provided we know that $u'_1(F)$ *and* $u'_2(F)$ are finite. However, the detours of going up from q to h and going down from h to q are responsible for a considerably weaker estimate than we might expect. For a general real field F (with $u'_2(F)$ finite) I can only prove an estimate of the type

$$
u'_r(F) \leq \text{const} \cdot r^6.
$$

Our treatment essentially follows the above-mentioned paper.

In this section the ground field is always *real* and is denoted by F. K and L will be finite extension fields of F which turn up later. V is an F-vectorspace with $\dim V = n < \infty$, $q = (q_1, \ldots, q_r) : V \to K^r$ is a quadratic map with $2 \times q \sim 0$, $b : V \times V \to K^r$ is the associated symmetric bilinear map defined by

$$
b(v, w) = \frac{1}{2}(q(v + w) - q(v) - q(w)).
$$

From Proposition 1.7 (3) we know that

$$w(F) = u(F(\sqrt{-1})) = u'(F(\sqrt{-1})) \le \frac{1}{2}u_2'(F).$$

This implies that $u_2'(F)$ and hence $u_r'(F)$ for $r \ge 2$ can only be finite if $w(F) < \infty$. Hence we make the *General Assumption*: $w = w(F) < \infty$.

From Theorem 2.12 of Chapter 8 we then get a good estimate for $u'(F) = u_1'(F)$, namely $u_1'(F) \le 2(w-1)$. On the other hand Examples 3.2 of Chapter 8 tell us that $u_1'(F) < \infty$ does not always imply $w(F) < \infty$.

A. Primary Decomposition

From Proposition 1.13 we know that for an *anisotropic* quadratic map $q : V \to F^r$ the condition $2 \times q \sim 0$ is equivalent to the existence of a linear transformation $T \in GL(V)$ with $q(Tv) = -q(v)$ for all $v \in V$. From this equation we get for all $v_1, v_2 \in V$ and $f(x) \in F[x]$

$$\begin{aligned} b(Tv_1, Tv_2) &= -b(v_1, v_2), \\ b(Tv_1, v_2) &= b(v_1, -T^{-1}v_2), \\ b(f(T)v_1, v_2) &= b(v_1, f(-T^{-1})v_2). \end{aligned}$$

Suppose $f(0) \ne 0$ and put $f^*(x) = f(0)^{-1}f(-x^{-1})x^{\deg f}$. Let $\mu = \mu_T$ be the minimal polynomial of T. Since T is regular we have $\mu(0) \ne 0$. Then μ^* is defined, it is monic and $\deg \mu^* = \deg \mu$. The equation

$$0 = b(\mu(T)v_1, v_2) = b(v_1, \mu(-T^{-1})v_2) = b(v_1, \mu^*(T)v_2)$$

for all $v_1, v_2 \in V$ now implies $\mu^*(T) = 0$, i.e. $\mu^* = \mu$ (since b is regular).

3.1 Lemma. Let $\mu = p_1 \ldots p_k$ be the (normed) prime factorization of μ. Then $\mu = \mu^* = p_1^* \ldots p_k^*$ and the p_i satisfy:

(1) $p_i^* = p_i$ for $i = 1, \ldots, k$, i.e. the p_i are "reciprocal".

(2) $p_i \ne p_j$ for $i \ne j$.

PROOF. Let $m_i = \deg p_i$. Put $W := (p_2 \ldots p_k)(T)V$. Then $W \ne 0$, W is T-invariant and $T|W$ has minimal polynomial p_1. We get for $w \in W$, $v \in V$

$$0 = b(p_1(T)w, T^{m_1}v) = b(w, p_1(-T^{-1})T^{m_1}v) = b(w, p_1^*(T)v).$$

(1) Assume $p_1^* \ne p_1$. Then there exists $v = w_1 \in W$ with $p_1^*(T)w_1 =: w \ne 0$. We get $0 = b(w, w) = q(w)$. This is a contradiction since q was supposed to be anisotropic.

(2) Assume $p_1 = p_2$. Choose $v \in (p_3 \ldots p_k)(T)V$ such that $w := p_1^*(T)v = p_1(T)v = p_2(T)v \neq 0$. Then $w \in W$ and again $b(w, w) = 0$: contradiction.

Property (2) in Lemma 3.1 together with char $F = 0$ shows that μ_T is a separable polynomial. Then $T \in GL(V)$ is semi-simple and every T-invariant subspace W of V has a T-invariant complement. Compare e.g. [Hoffmann–Kunze 1971, Ch. 7, Thm. 3 and Thm. 11].

3.2 Lemma. The prime factors p_i of μ have even degree $m_i = 2d_i$ and hence no zero in F.

PROOF. Since p_i is monic with $p_i^* = p_i$ we have

$$p_i^*(0) = p_i(0)^{-1}(-1)^{m_i} = p_i(0), \text{ i.e. } (-1)^{m_i} = p_i(0)^2.$$

Since $s(F) \neq 1$ this gives m_i even and $p_i(0) = \pm 1$.

For each $i \in \{1, \ldots, k\}$ put $q_i(x) = \frac{\mu(x)}{p_i(x)}$ and $V_i := q_i(T)V$. This is called the "primary component" of V w.r.t. p_i and T. $T_i := T|V_i$ has minimal polynomial p_i. From general linear algebra we know that V is the direct sum of the subspaces V_i ($i = 1, \ldots, k$). Furthermore we have

3.3 Lemma. The direct sum $V = V_1 \oplus \ldots \oplus V_k$ is an orthogonal sum w.r.t. the quadratic map q on V.

PROOF. Let $v_1 = q_1(T)v \in V_1$, $v_2 = q_2(T)v' \in V_2$ ($v, v' \in V$). Since $\mu | q_1 q_2$ we have $q_1^*(T)v_2 = q_1(T)v_2 = 0$. This implies $b(v_1, v_2) = b(v, q_1^*(T)v_2) = 0$.

It will turn out later that it is essentially enough to study the given system q on a fixed component V_i of V. In the next two sub-sections we shall therefore suppose that $\mu_T = p$ is a reciprocal prime polynomial of even degree $2d$. For the characteristic polynomial χ_T of T we then have $\chi_T = p^\ell$ for some $\ell \in \mathbb{N}$ and the dimension of V satisfies $n = \dim V = 2d\ell$.

B. Induced Quadratic Maps

For a reciprocal prime polynomial $p(x) \in F[x]$ two cases can occur:

(1) p has a zero $\alpha \in \bar{F}$ (the algebraic closure of F) with $\alpha = -\alpha^{-1}$. Then $\alpha^2 = -1$ and $p(x) = x^2 + 1$ since p is irreducible. As usual we put $\alpha = \sqrt{-1} =: i$.

(2) For each zero $\alpha \in \bar{F}$ of p we have $\alpha \neq -\alpha^{-1}$. This second case is more complicated, it will be treated in sub-section C.

Let us now assume that $\mu_T(x) = x^2 + 1$ for our automorphism $T \in GL(V)$ with $q(Tv) = -q(v)$. Then $\chi_T(x) = (x^2 + 1)^\ell$ for some $\ell \in \mathbb{N}$, hence $n = \dim V = 2\ell$. Thus V may be considered as an ℓ-dimensional vector-space over $F(i)$. For each $F(i)$-quadratic map

$$\tilde{q} : V \to F(i)^r \cong F^r \oplus iF^r$$

its real part Re $\tilde{q}$ and its imaginary part Im $\tilde{q}$ are F-quadratic maps from V to F^r.

3.4 Definition. A quadratic map $q : V \to F^r$ is called *induced* (from $F(i)$) if there is an $F(i)$-quadratic map $\tilde{q} : V \to F(i)^r$ such that $q = $ Im $\tilde{q}$.

The main result of this sub-section B reads as follows:

3.5 Theorem. Let V be an F-vector-space of even dimension 2ℓ. Let $q : V \to F^r$ be a quadratic map. Then q is induced if and only if there exists $T \in GL(V)$ with $T^2 = -1$ and $q(Tv) = -q(v)$ for all $v \in V$. In particular, every induced map q satisfies $2 \times q \sim 0$.

PROOF.

(1) Let $q = $ Im $\tilde{q}$ be induced. Define $T \in GL(V)$ by $Tv := iv$ $(v \in V)$. Then T is F-linear with $T^2 = -1$. Since $\tilde{q}(iv) = i^2\tilde{q}(v) = -\tilde{q}(v)$, we also have $q(Tv) = -q(v)$.

(2) Conversely assume $T^2 = -1$, $q(Tv) = -q(v)$. The 2ℓ-dimensional F-vector-space V is turned into an ℓ-dimensional $F(i)$-vector-space if we define $iv := Tv$. Define

$$\tilde{b}(v_1, v_2) := b(Tv_1, v_2) + ib(v_1, v_2) \text{ for } v_1, v_2 \in V$$

where b is the associated bilinear form of q. $\tilde{b}$ is clearly F-bilinear. Furthermore we have

$$\tilde{b}(iv_1, v_2) = \tilde{b}(Tv_1, v_2) = b(T^2v_1, v_2) + ib(Tv_1, v_2) = i\tilde{b}(v_1, v_2)$$

and

$$b(Tv_1, v_2) = b(Tv_1, T(T^{-1}v_2)) = -b(v_1, T^{-1}v_2) = b(v_1, Tv_2) = b(Tv_2, v_1).$$

Hence $\tilde{b}$ is $F(i)$-bilinear and symmetric. The map $\tilde{q}(v) := \tilde{b}(v, v)$ is $F(i)$-quadratic, $\tilde{q} : V \to F(i)^r$, and it satisfies Im $\tilde{q} = q$. Hence q is induced.

3.6 Note. From Proposition 2.9 and Theorem 2.12, in Chapter 8, we know that *every* (regular) quadratic form q_1 with $2 \times q_1 \sim 0$ over a field F with $s(F) \neq 1$ is induced from $F(i)$. But this does not generalize to systems $q = $

$(q_1, \ldots, q_r)$ with $2 \times q \sim 0$ if $r \geq 2$, at least not for $r \geq 3$. In general the existence of a transformation $T \in GL(V)$ with $q(Tv) = -q(v)$ does not imply the existence of such a transformation with the *additional property* $T^2 = -1$. This will be pretty clear from the next sub-section C (though I omit explicit examples).

Induced anisotropic maps $q : V \to F^r$ are easy to handle:

3.7 Definition. Let F be real with finite invariant $w = w(F) = u(F(i))$. Put $w_r = w_r(F) := \sup\{\dim V$: there exists an anisotropic induced quadratic map $q = (q_1, \ldots, q_r) : V \to F^r\}$.

With this definition we have

3.8 Theorem. $w_r \leq r[(r + 1)w - 2]$ for all $r \in \mathbb{N}$.

PROOF. Let $q : V \to F^r$ be anisotropic and induced. Then $q = \text{Im } \tilde{q}$ where $\tilde{q} : V \to F(i)^r$ is anisotropic, or more precisely: For $0 \neq v \in V$ not all components $\tilde{q}_\varrho(v)$ of $\tilde{q}(v)$ ($\varrho = 1, \ldots, r$) lie in F. From Theorem 2.1 we know that $\dim_{F(i)} V \leq \frac{r(r+1)}{2}w$ which gives the a priori estimate $w_r \leq r(r+1)w$. But if we carefully check the proof of 2.1 we see that $\tilde{q}_1$ cannot be universal on a (maximal) $F(i)$-subspace U of V with $(\tilde{q}_2, \ldots, \tilde{q}_r)|U \equiv 0$. Hence $\dim_{F(i)} U \leq w - 1$. The estimates $\dim_{F(i)} W_0 \leq \frac{r(r-1)}{2}w$ and $\dim_{F(i)} V \leq r \dim_{F(i)} U + \dim_{F(i)} W_0$ remain valid. We obtain

$$\dim_F V = 2 \dim_{F(i)} V \leq 2r(w - 1) + r(r - 1)w = r((r + 1)w - 2).$$

For $r = 1$ the estimate $w_1 \leq 2(w - 1)$ coincides with the estimate $u'(F) \leq 2(w - 1)$ of Chapter 8, Theorem 2.12.

3.9 Corollary. Let $F = R$ be real-closed. Then $u'_r(F) \leq 2(r - 1)$.

PROOF. Let $q : V \to F^r$ be anisotropic with $2 \times q \sim 0$. Let $T \in GL(V)$ be such that $q(Tv) = -q(v)$ for all $v \in V$, and let $\mu_T = p_1 \ldots p_k$ be the minimal polynomial of T. By sub-section A the p_i are different reciprocal prime polynomials. But Lemma 3.2 shows that $p_1(x) = x^2 + 1$ is the only reciprocal prime polynomial if F is real-closed. Hence $k = 1$ and $\mu_T = p_1$, i.e. $T^2 = -1$. Then 3.5 implies that q is induced. Finally $C = F(i)$ is algebraically closed, hence $u_{r-1}(C) = r - 1$ by Hilbert's nullstellensatz. Then $U = 0$ and $\dim_{F(i)} W_0 = r - 1$ in the above proof. This gives the desired result $\dim_F V \leq 2(r - 1)$.

C. Hermitian Maps

We come to the second case for a reciprocal prime polynomial p of degree $2d$. Here $\lambda \neq -\lambda^{-1}$ for every root λ of p. Then the zeros $\lambda_1, \ldots, \lambda_{2d}$ of p can be indexed in such a way that

$$\lambda_{2i-1} = t_i, \quad \lambda_{2i} = -t_i^{-1} \quad (i = 1, \ldots, d), \ t_i \in \bar{F}.$$

Put $t_i - t_i^{-1} =: 2u_i$. Then we have in $\bar{F}[x]$

$$p(x) = \prod_{i=1}^{d}(x^2 - 2u_i x - 1).$$

For brevity put $u = u_1$, $t = t_1$, $K = F(u)$, $L = F(t)$. Then $[K : F] = d$, $[L : K] = 2$, $L = K(t) = K(\sqrt{1 + u^2})$ since p is irreducible over F. We write the elements of L in the form $\alpha = x_1 + x_2 t$ with $x_1, x_2 \in K$ and identify L with K^2 (as a K-vector-space). $t \to \bar{t} := -t^{-1}$ is the nontrivial automorphism of L/K. Using a fixed basis $1 = w_1, \ldots, w_d$ of K/F we can further identify the following F-vector-spaces: $K = F^d$, $L = K^2 = F^{2d}$. The F-linear map $\alpha \to t\alpha$ ($\alpha \in L$) is then identified with an endomorphism T_0 of F^{2d}. Clearly T_0 has the same minimal polynomial over F as t, namely $p(x)$. The original transformation $T \in GL(V)$ whose characteristic polynomial was $\chi_T = p^\ell$ is then similar to the ℓ-fold direct sum of T_0 (use that T is semi-simple). But a similarity $S \in GL(V)$ takes T to $S^{-1}TS$ and q to an equivalent quadratic map q' with $q'(v) := q(Sv)$. Hereby the properties "q anisotropic, $2 \times q \sim 0$" remain unchanged. This allows us to assume w.l.o.g. that

$$V = F^{2d\ell} = L^\ell,$$

$T \in GL(V)$ is the scalar multiplication by t on L^ℓ, i.e.

$$T \begin{pmatrix} \alpha_1 \\ \vdots \\ \alpha_\ell \end{pmatrix} = \begin{pmatrix} t\alpha_1 \\ \vdots \\ t\alpha_\ell \end{pmatrix}.$$

Next we will show that the components q_ϱ of our quadratic map $q = (q_1, \ldots, q_r)$ with $q(Tv) = -q(v)$ are induced from ℓ-dimensional *hermitian* forms h_ϱ over L (where $\alpha^* := \bar{\alpha}$ for $\alpha \in L$, $\lambda := 1$). To do this we start by counting dimensions. Let for the moment $q : V \to F$ denote a single quadratic form with $q(Tv) = -q(v)$ for all v. Relative to a *fixed* basis of V we have $q(v) = v'Av$ where $A \in M_n(F)$ is a symmetric matrix with $T'AT = -A$. Put $\mathcal{A} = \{A \in M_n(F) : A = A', T'AT = -A\}$. Similarly a hermitian form h on L^ℓ corresponds to a hermitian matrix $H = \bar{H}' \in M_\ell(L)$ such that $h(\alpha, \beta) = \alpha'H\bar{\beta}$ for $\alpha, \beta \in L^\ell$. Put $\mathcal{H} = \{H \in M_\ell(L) : H = \bar{H}'\}$.

3.10 Lemma. $\mathcal{A}$ and $\mathcal{H}$ are F-vector-spaces of the same dimension $\dim \mathcal{A} = \dim \mathcal{H} = d\ell^2$.

PROOF.

(1) For $\mathcal{H}$ the proof is nearly trivial. If $H = (\eta_{ij})_{i,j=1,\ldots,\ell}$ then the elements η_{ij} above the diagonal (i.e. for $i < j$) are arbitrary in L whereas the elements $\eta_{ii} = \bar{\eta}_{ii}$ on the diagonal are arbitrary in K, and $\eta_{ji} = \bar{\eta}_{ij}$ for $i < j$. This shows that $\mathcal{H}$ is a K-vector-space of dimension $\ell + 2\frac{\ell(\ell-1)}{2} = \ell^2$. Thus $\dim_F \mathcal{H} = d\ell^2$.

(2) For $\mathcal{A}$ the equations $A = A'$, $T'AT = -A$ lead to a system of F-linear equations for the coefficients a_{ij} of A $(i, j = 1, \ldots, n;\ n = 2d\ell)$. Let us solve this system over $\bar{F}$. Then T_0 is similar to the $2d$-dimensional diagonal matrix with entries λ_i $(i = 1, \ldots, 2d)$ and T is similar to the block diagonal matrix $(\lambda_i I_\ell)_{i=1,\ldots,2d}$ where I_ℓ is the $\ell \times \ell$ unit matrix. Correspondingly A has block form $A = (A_{jk})_{j,k=1,\ldots,2d}$. We have to solve the equations $A = A'$ and $\lambda_j A_{jk} \lambda_k = -A_{jk}$ over $\bar{F}$. This implies $A_{jk} = 0$ if $\lambda_j \lambda_k \neq -1$, A_{jk} arbitrary if $\lambda_j \lambda_k = -1$, $A_{kj} = A'_{jk}$. This means that the matrices $A_{2i-1,2i}$ $(i = 1, \ldots, d)$ are arbitrary $\ell \times \ell$ matrices over $\bar{F}$, that $A_{2i,2i-1} = A'_{2i-1,2i}$ and that $A_{jk} = 0$ otherwise. Thus over $\bar{F}$ the corresponding system of linear equations has solution space of dimension $d\ell^2$. Hence $\dim \mathcal{A} = d\ell^2$.

The main point of our investigation is now that there is a natural injective (and hence surjective) F-linear map $\mathcal{H} \to \mathcal{A}$. This is seen as follows:

a) Let $h(\alpha, \beta) = \alpha' H \bar{\beta}$ be a hermitian form on L^ℓ $(H \in \mathcal{H})$. It induces a K-quadratic form $Q_h(\alpha) := h(\alpha, \alpha)$ on $L^\ell = K^{2\ell}$ since K is fixed under the automorphism $^-$.

If h is regular then Q_h is also regular: Let B_h be the symmetric bilinear form associated to Q_h over K. Assume $B_h(\alpha_1, \alpha_2) = 0$ for all $\alpha_1 \in L^\ell = K^{2\ell}$. Since $B_h(\alpha_1, \alpha_2) = \frac{1}{2}(h(\alpha_1, \alpha_2) + \overline{h(\alpha_1, \alpha_2)})$ the element $\beta = h(\alpha_1, \alpha_2) \in L$ satisfies $\bar{\beta} = -\beta$ (for all α_1). Replace α_1 by $t\alpha_1$. Then we also have $\bar{t}\bar{\beta} = -t\beta$, $-t^{-1}(-\beta) = -t\beta$, $-\beta = t^2\beta$. But $t^2 \neq -1$ (since we are in the second case), hence $\beta = 0$ for all $\alpha_1 \in L$. This implies $\alpha_2 = 0$ if h is regular. Similarly, if h has rank $m \leq \ell$ over L then Q_h has rank $2m$ over K. This is best seen by diagonalizing h. (Compare [S, Ch. 7, Thm. 6.3].)

b) Let $s : K \to F$ be a *fixed* nontrivial F-linear map, e.g. $s(1) = 1$, $s(w_i) = 0$ for $i = 2, \ldots, d$. Then the Scharlau transfer s_* maps the K-quadratic form Q_h to an F-quadratic form $q_h := s_*(Q_h)$. Here again it is easy to see that for Q_h of rank $2m$ the induced form q_h has rank $2dm$. (See [S, Ch. 2, Lemma 5.5].)

c) The equation $t\bar{t} = -1$ in L implies $Q_h(t\alpha) = h(t\alpha, t\alpha) = th(\alpha, \alpha)\bar{t} = -Q_h(\alpha)$ for all $\alpha \in L^\ell$ and all h (resp. H) $\in \mathcal{H}$. If we apply s_* and remember the identification $V = L^\ell$ we get $q_h(Tv) = -q_h(v)$ for all $v \in V$. Let A denote the symmetric $n \times n$ matrix of q_h. We have proved

3.11 Theorem. The map $\mathcal{H} \to \mathcal{A}$ which sends h to $q_h = s_*(Q_h)$ and H to A is F-linear and bijective. Every quadratic form q on V which satisfies $q(Tv) = -q(v)$ is of the form $q = q_h$ for a uniquely determined hermitian form h on L^ℓ.

The main advantage of replacing $q = q_h$ by h is the following: The condition $q(Tv) = -q(v)$ is replaced by the condition $t\bar{t} = -1$ which is a property of the field L and the element t alone. The hermitian forms h on L^ℓ do not have to satisfy any side conditions. In particular they can be restricted to any L-subspace of L^ℓ and Theorem 2.5 applies for them. The disadvantage of this procedure is, however, that in going up from q to h and going back from h to q we diminish the strength of the final estimate for $u_r'(F)$. On the other hand, Theorem 3.11 also shows that for every pair of extension fields $K = F(u)$, $L = K(\sqrt{1 + u^2}) \neq K$ and every hermitian form h on L^ℓ the induced quadratic form q_h over F satisfies $2 \times q_h \sim 0$. Therefore the hermitian forms h are so intimately connected to the quadratic forms q with $2 \times q \sim 0$ that it is unlikely we could avoid them when studying $u_r'(F)$.

We can now go back to the given anisotropic system $q = (q_1, \ldots, q_r)$ on V with $q(Tv) = -q(v)$. By 3.11 we have $q_\varrho = q_{h_\varrho}$ ($\varrho = 1, \ldots, r$) with $h_\varrho \in \mathcal{H}$. Clearly the system $h := (h_1, \ldots, h_r)$ must be anisotropic on L^ℓ. By Theorem 2.5 this implies $\ell \leq u_r(L, \bar{\,}, 1) \leq \frac{r(r+1)}{1} u_1(L, \bar{\,}, 1)$. The invariant $u_1(L, \bar{\,}, 1)$ is finite. We have

3.12 Proposition.

(1) $2u_1(L, \bar{\,}, 1) \leq u_1'(K)$.

(2) $u_1'(K) \leq (d + 1)w$ (with $w = w(F)$).

PROOF.

(1) Let h be an anisotropic hermitian form over L. Then (by definition) the induced quadratic form Q_h over K is anisotropic. It satisfies $Q_h(t\alpha) = -Q_h(\alpha)$ for all α, hence $2 \times Q_h \sim 0$. This implies $2 \dim h = \dim Q_h \leq u_1'(K)$.

(2) From previous estimates we get

$$u_1'(K) \leq 2u_1'(K(i)) \leq \frac{2}{d} u_d'(F(i)) \leq (d + 1)u_1'(F(i)) = (d + 1)w.$$

An interesting special case occurs if the hermitian matrices H_ϱ of the forms h_ϱ have all their entries in K (for $\varrho = 1, \ldots, r$). Then it is possible to replace the transformation $\alpha \to t\alpha$ ($\alpha \in L^\ell$) by $\alpha \to t\bar{\alpha}$ because we have

$$Q_\varrho(t\bar{\alpha}) = h_\varrho(t\bar{\alpha}, t\bar{\alpha}) = t\bar{\alpha}' H_\varrho \bar{t}\alpha = t\bar{t}\alpha' H_\varrho' \bar{\alpha} = -\alpha' H_\varrho \alpha = -q_\varrho(\alpha)$$

if $H_\varrho' = \bar{H}_\varrho = H_\varrho$. Let $\tilde{T}$ be the F-linear transformation of V which is induced from the F-linear transformation $\alpha \to t\bar{\alpha}$ of L^ℓ. Then $q_\varrho(\tilde{T}v) = -q_\varrho(v)$ for all $v \in V$ and $\tilde{T}^2 = -1$ since $t(\overline{t\bar{\alpha}}) = t\bar{t}\alpha = -\alpha$. This shows:

3.13 Proposition. If the matrices H_ϱ ($\varrho = 1, \ldots, r$) are K-rational then the system $q = (q_1, \ldots, q_r)$ is induced from $F(i)$.

For $r = 1$ we can always assume this favourite case since every hermitian matrix H_1 can be diagonalized, and the entries on the diagonal of H_1 are automatically in K. This gives another proof of the results 2.9, 2.12 of Chapter 8.

D. Estimates for $u'_r(F)$

By collecting the results of sub-sections A – C we are now able to prove an estimate for $u'_r(F)$ depending only on r and the invariant $w = w(F)$.

Let $q : V \to F^r$ be an anisotropic quadratic map with $2 \times q \sim 0$, let $\dim V = n$. From sub-section A we know that there exists $T \in GL(V)$ with $q(Tv) = -q(v)$ for all $v \in V$ and that T has characteristic polynomial $\chi_T = p_1^{\ell_1} \ldots p_k^{\ell_k}$ where the p_j are (different) reciprocal prime polynomials of even degree $2d_j$. Let K_j, L_j be the extension fields of sub-section C defined by p_j. In the "exceptional" case $p_j(x) = x^2 + 1$ we have to put $K_j = F$, $L_j = F(i)$. V has the orthogonal primary decomposition $V = V_1 \oplus \ldots \oplus V_k$ with respect to q and we have $n = 2d_1\ell_1 + \ldots + 2d_k\ell_k$. The quadratic maps $q_j := q|V_j$ are of course anisotropic. In the exceptional case we have $q_j = \text{Im } \tilde{q}_j$ where $\tilde{q}_j$ is an ℓ_j-dimensional $F(i)$-quadratic map $V_j \to F(i)^r$ (by sub-section B), otherwise we have $q_j = s_{j*}(q_{h_j})$ with a hermitian map $h_j : L_j^{\ell_j} \to L_j^r$ (by sub-section C). The main result reads as follows:

3.14 Theorem. For every real field F and every $r \in \mathbb{N}$ we have

$$u'_r(F) \leq \left(\frac{r(r+1)w}{2}\right)^2 \left(\frac{r(r+1)w}{2} + 1\right).$$

For $r \geq 2$ this estimate can be weakened to $u'_r(F) \leq \frac{9}{16}r^6 w^3$.

PROOF.

(1) Our first aim is to estimate $2d := 2\sum_{j=1}^{k} d_j = \deg \mu_T$. This can be done by restricting each hermitian map h_j to an arbitrary 1-dimensional subspace of $L_j^{\ell_j}$. Let $\bar{h}_j$ denote this restricted map. Then Proposition 3.13 applies. The quadratic map $\bar{q}_j = s_{j*}(q_{\bar{h}_j}) : U_j \to F^r$ is induced from $F(i)$, U_j is a subspace of dimension $2d_j$ of V_j. Clearly $\bar{q} := \bar{q}_1 \oplus \ldots \oplus \bar{q}_k$ is anisotropic since it is the restriction of q to the subspace $U = U_1 \oplus \ldots \oplus U_k$. (In the exceptional case $q_j = \text{Im } \tilde{q}_j$ itself is induced. Here we can restrict $\tilde{q}_j$ to a 1-dimensional subspace.) This shows that $\bar{q}$ is anisotropic and induced from $F(i)$. From Theorem 3.8 we get $2d = \dim U \leq w_r \leq r(r+1)w$.

(2) For the multiplicities ℓ_j we have from Theorem 2.5 and Proposition 3.12

$$\ell_j \le \frac{r(r+1)}{2} u_1(L_j,^-,1) \le \frac{r(r+1)}{4}(d_j+1)w.$$

(In the exceptional case we have from Theorem 3.8 dim $V_j = 2d_j\ell_j = 2\ell_j \le w_r \le r(r+1)w$. This amounts to the same estimate for ℓ_j as above.)

(3) Summing up the estimates (2) for $j = 1, \ldots, k$ we get

$$n = \sum_{j=1}^{k} 2d_j\ell_j \le \frac{r(r+1)w}{2} \sum_{j=1}^{k} d_j(d_j+1)$$

with the side condition $\sum_{j=1}^{k} d_j \le \frac{r(r+1)w}{2}$ from (1).

(4) By induction on k it is easily seen that

$$\sum_{j=1}^{k} d_j \le c \quad \text{implies} \quad \sum_{j=1}^{k} d_j(d_j+1) \le c(c+1).$$

Hence we get from (3)

$$n \le \frac{r(r+1)w}{2} \cdot \frac{r(r+1)w}{2} \left(\frac{r(r+1)w}{2} + 1 \right).$$

This implies the estimate for $u'_r(F)$. The simpler estimate for $r \ge 2$ follows from $\frac{r+1}{2} \le \frac{3}{4}r$ and $1 \le \frac{r^2w}{4}$ (for $r \ge 2$).

In many examples the bound for u'_r can be considerably improved. This is the case if there is a *common* bound for the w-invariants of F and all its algebraic extension fields which gives an improvement of estimate (2), or if $u_r(F(i))$ has a better bound than the one coming from Leep's Theorem 2.1. In this case (1) can be improved as well.

3.15 Corollary. Let F be a (real) number field. Then

$$u'_r(F) \le (2r(r+1))^2.$$

PROOF. We have $w = 4$ and $u'_1(K_j) = 4$ for all j. This gives $u_1(L_j,^-,1) \le 2$ in the non-exceptional case and $\ell_j \le r(r+1)$ in all cases.
Hence

$$n = \sum_{j=1}^{k} 2d_j\ell_j = r(r+1) \sum_{j=1}^{k} (2d_j) \le r(r+1) \cdot 4r(r+1) \quad \text{by (1)}.$$

3.16 Corollary. Let F be a (real) field of transcendence degree $m \geq 0$ over a real-closed field R. Then

$$u'_r(F) \leq 4^m \cdot r^2(r+1).$$

PROOF. Here (1) can be replaced by

(1″)

$$\sum_{j=1}^{k}(2d_j) = 2d \leq 2\dim_{F(i)} U \leq 2 \cdot 2^m \cdot r$$

since $F(i)$ is a C_m-field. (2) can be replaced by

(2″)

$$\ell_j \leq \frac{r(r+1)}{4}u'_1(K_j) \leq \frac{r(r+1)}{4} \cdot 2u_1(K_j(i)) \leq \frac{r(r+1)}{2} \cdot 2^m$$

since $K_j(i)$ is a C_m-field. Hence

$$n = \sum_{j=1}^{k} 2d_j\ell_j \leq \frac{r(r+1)}{2}2^m \cdot \sum_{j=1}^{k}(2d_j) \leq 4^m r^2(r+1).$$

Chapter 10

The Level of Projective Spaces

§1. Algebraic and Topological Preliminaries

In the last chapter we have seen that "induced" systems of quadratic forms play a decisive role in the investigation of the system-u-invariant $u'_r(F)$ of a formally real field F. Here we are concerned with the special field $F = \mathbf{R}$ where automatically every anisotropic system $q = (q_1, \ldots, q_r)$ with $2 \times q \sim 0$ is induced since $\mathbf{C} = \mathbf{R}(i)$ is the only algebraic extension field of $\mathbf{R}$. Therefore we have

$u'_r(\mathbf{R}) = w_r(\mathbf{R}) = \sup\{2m :$ there exists a system $\varphi = (\varphi_1, \ldots, \varphi_r)$ over $\mathbf{C}$ in m complex variables such that $q = \text{Im } \varphi$ is anisotropic $\}$.

As the fields $\mathbf{R}, \mathbf{C}$ are fixed we introduce a new invariant $m(r)$ as follows:

1.1 Notation. For $r \in \mathbf{N}$ put

$m(r) := \sup\{m :$ There exists a quadratic map $\varphi : \mathbf{C}^m \to \mathbf{C}^r$ such that the induced map $q = \text{Im } \varphi : \mathbf{R}^{2m} \to \mathbf{R}^r$ is anisotropic$\}$

$\quad = \sup\{m :$ there exists $\varphi : \mathbf{C}^m \to \mathbf{C}^r$ such that $\varphi(\mathbf{C}^m \backslash 0) \subset \mathbf{C}^r \backslash \mathbf{R}^r\}$.

Clearly we have $m(r) \in \mathbf{N}_0 \cup \infty$ and $u'_r(\mathbf{R}) = 2m(r)$. We would like to compute $m(r)$ for all $r \in \mathbf{N}$. It will turn out that this is much more difficult than expected at first sight and that this purely algebraic invariant $m(r)$ is intimately related to a topological invariant $s(m)$. The final result will be that we have some easy explicit estimates for $m(r)$ from below and above and some good conjectures for the precise value of $m(r)$.

Let us start with some elementary estimates for $m(r)$.

1.2 Proposition. $m(r) \leq r - 1$ for all $r \in \mathbf{N}$.

PROOF. Let $\varphi = (\varphi_1, \ldots, \varphi_r)$ be a quadratic map from $\mathbf{C}^m$ to $\mathbf{C}^r$ where $m \geq r$. By Hilbert's homogeneous nullstellensatz the system of equations

$$\varphi_1(z) = \ldots = \varphi_{r-1}(z) = 0$$

has a nontrivial solution $0 \neq z_0 \in \mathbf{C}^m$ (since $m > r - 1$). Now look at $\varphi_r(z_0)$. This is some complex number. We can find $t \in \mathbf{C}$, $|t| = 1$ such that $\varphi_r(tz_0) = t^2 \varphi_r(z_0) = |\varphi_r(z_0)| \in \mathbf{R}$. Then

$$\varphi(tz_0) = (0, \ldots, 0, |\varphi_r(z_0)|) \in \mathbf{R}^r,$$

i.e. Im φ is isotropic whenever $m \geq r$. Hence $m(r) \leq r - 1$. Compare Chapter 9, Corollary 3.9.

1.3 Proposition. $m(r)$ is a non-decreasing function of r,

$$m(r_1 + r_2) \geq m(r_1) + m(r_2).$$

PROOF. For a system $\varphi = (\varphi_1, \ldots, \varphi_{r+1})$ let $\varphi' = (\varphi_1, \ldots, \varphi_r)$ be the reduced system by omitting the last component. If $\varphi'(\mathbf{C}^m \backslash 0) \subset \mathbf{C}^r \backslash \mathbf{R}^r$ then clearly $\varphi(\mathbf{C}^m \backslash 0) \subset \mathbf{C}^{r+1} \backslash \mathbf{R}^{r+1}$. This implies $m(r + 1) \geq m(r)$.

Let now $\varphi = (\varphi_1, \ldots, \varphi_{r_1})$, $\psi = (\psi_1, \ldots, \psi_{r_2})$ be systems in $m_1 = m(r_1)$ and $m_2 = m(r_2)$ *disjoint* complex variables respectively such that

$$\varphi(\mathbf{C}^{m_1} \backslash 0) \subset \mathbf{C}^{r_1} \backslash \mathbf{R}^{r_1}, \quad \psi(\mathbf{C}^{m_2} \backslash 0) \subset \mathbf{C}^{r_2} \backslash \mathbf{R}^{r_2}.$$

Put $\chi = (\varphi, \psi) = (\varphi_1, \ldots, \varphi_{r_1}, \psi_1, \ldots, \psi_{r_2})$ which is a system of $r_1 + r_2$ quadratic forms in $m_1 + m_2$ variables. We have $\chi(\mathbf{C}^{m_1 + m_2} \backslash 0) \subset \mathbf{C}^{r_1 + r_2} \backslash \mathbf{R}^{r_1 + r_2}$. Hence $m(r_1 + r_2) \geq m_1 + m_2$.

1.4 Proposition. We have $m(1) = 0$, $m(2) = 1$, $m(3) = 2$.

PROOF. From 1.2 we trivially get $m(1) = 0$, $m(2) \leq 1$, $m(3) \leq 2$. The system $\varphi = (z_1^2, iz_1^2) : \mathbf{C} \to \mathbf{C}^2$ assumes no real value unless $z_1 = 0$. This shows $m(2) = 1$. For $r = 3$ consider the system $\varphi = (\varphi_1, \varphi_2, \varphi_3) : \mathbf{C}^2 \to \mathbf{C}^3$ with $z = (z_1, z_2)$, $\varphi_1(z) = 2z_1 z_2$, $\varphi_2(z) = z_1^2 - z_2^2$, $\varphi_3(z) = i(z_1^2 + z_2^2)$. Then $\varphi_1^2 + \varphi_2^2 + \varphi_3^2 = 0$ identically in z. Let $z \in \mathbf{C}^2$ be such that $\varphi(z) \in \mathbf{R}^3$. This can only hold if $\varphi_1(z) = \varphi_2(z) = \varphi_3(z) = 0$ which implies $z = (0, 0)$. Hence $m(3) \geq 2$ by the definition of $m(r)$.

Combining 1.2, 1.3 and 1.4 we get

1.5 Proposition. $\left[\frac{2r}{3}\right] \leq m(r) \leq r - 1$ for all $r \in \mathbf{N}$.

For small $r > 3$ this leaves the possibilities $m(4) = 2$ or 3, $m(5) = 3$ or 4, $m(6) = 4$ or 5, $m(7) = 4$ or 5 or 6 etc. It seems plausible that the upper bound $r - 1$ for $m(r)$ is pretty sharp since it follows from Hilbert's nullstellensatz which cannot be improved. On the other hand an improvement of the lower bound $\left[\frac{2r}{3}\right]$ would require the construction of quadratic maps $\varphi = (\varphi_1, \ldots, \varphi_r)$ of "large" dimension m and "without real values" (see the definition of $m(r)$). This might well be possible on the basis of a clever idea.

From our further results it follows that we have $m(r) \leq r - 2$ for even numbers $r \geq 4$, in particular $m(4) = 2$, $m(6) = 4$. But this involves hard topology. Therefore we ask the following.

1.6 Open Question. Is there an "easy" proof for the fact that $m(r) \leq r-2$ for even $r \geq 4$? Easy means that it should only involve algebraic methods and perhaps some elementary results from algebraic topology.

Let us now interchange the roles of the number r of forms and the number m of complex variables, i.e. m is given and r is considered as a function of m. This leads to the following.

1.7 Notation. For $m \in \mathbb{N}$ put
$$r(m) := \min\{r : \text{there exists a quadratic map } \varphi : \mathbb{C}^m \to \mathbb{C}^r \text{ such that}$$
$$\text{Im } \varphi : \mathbb{R}^{2m} \to \mathbb{R}^r \text{ is anisotropic}\}$$
$$= \min\{r : \text{there exists } \varphi : \mathbb{C}^m \to \mathbb{C}^r \text{ such that } \varphi(\mathbb{C}^m \backslash 0) \subset \mathbb{C}^r \backslash \mathbb{R}^r\}.$$

1.8 Proposition. $r(m)$ is a non-decreasing function of m,

$$r(m_1 + m_2) \leq r(m_1) + r(m_2).$$

PROOF. If $\varphi : \mathbb{C}^m \to \mathbb{C}^r$ is such that Im φ is anisotropic then $\tilde{\varphi} = \varphi(z_1, \ldots, z_{m-1}, 0) : \mathbb{C}^{m-1} \to \mathbb{C}^r$ has the same property, hence $r(m-1) \leq r(m)$. The second statement follows from the composition of two quadratic maps $\varphi_1 : \mathbb{C}^{m_1} \to \mathbb{C}^{r_1}$, $\varphi_2 : \mathbb{C}^{m_2} \to \mathbb{C}^{r_2}$ in disjoint sets of variables: $\varphi = (\varphi_1, \varphi_2) : \mathbb{C}^{m_1 + m_2} \to \mathbb{C}^{r_1 + r_2}$.

Proposition 1.5 implies the following estimates for $r(m)$:

1.9 Theorem. $m + 1 \leq r(m) \leq \frac{1}{2}(3m + 1)$.

In particular we have $r(1) = 2$, $r(2) = 3$, $4 \leq r(3) \leq 5$, $5 \leq r(4) \leq 6$, $6 \leq r(5) \leq 8$ etc.

The correspondence between the functions $m(r)$, $r(m)$ can be expressed as follows:

1.10 Proposition. Let m, r be natural numbers. Then the following statements are equivalent:

(1) For *every* complex quadratic map $\varphi : \mathbb{C}^m \to \mathbb{C}^r$ the induced map $q = $ Im $\varphi : \mathbb{R}^{2m} \to \mathbb{R}^r$ is isotropic.

(2) $m > m(r)$.

(3) $r < r(m)$.

PROOF. Trivial from the definitions of $m(r)$, $r(m)$.

As an immediate consequence we have
1.11 Corollary.

(1) If the function $m(r)$ is known then

$$r(m) - 1 = \max\{r : m(r) < m\}.$$

(2) If the function $r(m)$ is known then

$$m(r) + 1 = \min\{m : r(m) > r\}.$$

(3) $m(r) \leq r - 2$ for even $r > 2 \iff r(m) \geq m + 2$ for odd $m > 1$.

(It will be shown in section 2 that the latter inequality is true.)

We shall now turn to the topological consequences of the existence of an anisotropic induced quadratic map $q = \operatorname{Im} \varphi : \mathbf{R}^{2m} \to \mathbf{R}^r$ for suitable numbers m, r (i.e. $r \geq r(m)$ or $m \leq m(r)$). Obviously such a map q induces a continuous map $\tilde{f} : S^{2m-1} \to S^{r-1}$ defined by $\tilde{f} := \frac{q(v)}{\|q(v)\|}$ for $v \in S^{2m-1} \subset \mathbf{R}^{2m}$, $q(v) \in \mathbf{R}^r \backslash 0$. Furthermore we see that q induces an equivariant map $f : \mathbf{R}P^{2m-1} \multimap S^{r-1}$: Denote the points of $\mathbf{C}^m = \mathbf{R}^{2m}$ by $z = (z_1, \ldots, z_m)$ with $z_j = x_j + iy_j \in \mathbf{C}$, $x_j, y_j \in \mathbf{R}$ $(j = 1, \ldots, m)$. Then the points of the $(2m - 1)$-dimensional real projective space $\mathbf{R}P^{2m-1}$ are given as equivalence classes $[z] = z \bmod \sim$ where $z \sim z'$ iff $z = \lambda z'$ for some $0 \neq \lambda \in \mathbf{R}$ $(z \neq 0)$. Multiplication by the imaginary unit i induces a self-map $z \to iz$ of $\mathbf{C}^m \backslash 0$ and a self-map $[z] \to [iz]$ of $\mathbf{R}P^{2m-1}$, both without fixpoints. Since $[i^2 z] = [-z] = [z]$ the self-map of $\mathbf{R}P^{2m-1}$ is an involution which will be denoted by i again (without too much danger of confusion). For the quadratic map q we have $q(-z) = q(z)$ and $q(iz) = i^2 q(z) = -q(z)$, hence $\tilde{f}(iz) = -\tilde{f}(z)$. Finally $\tilde{f}$ induces a map

$$
\begin{array}{ccccccc}
f & : & \mathbf{R}P^{2m-1} & = & \mathbf{C}^{2m}\backslash 0 \Big/ \mathbf{R}^\times & = & S^{2m-1}\Big/\pm & \to & S^{r-1} \\
& & & & & & [z] & \longmapsto & \frac{q(z)}{\|q(z)\|}
\end{array}
$$

which is equivariant with respect to the involutions i on $\mathbf{R}P^{2m-1}$ and the antipodal involution $-$ on S^{r-1}.

This brings us back to the theme of Chapter 3, namely the level of topological spaces with involution. We use the following.

1.12 Notation. For $m \in \mathbf{N}$ the level of the projective space $\mathbf{R}P^{2m-1}$ with its natural involution i is denoted by

$$
s(m) := s(\mathbf{R}P^{2m-1}, i).
$$

The above considerations give

1.13 Proposition. Every induced anisotropic quadratic map $q = \operatorname{Im} \varphi : \mathbf{R}^{2m} \to \mathbf{R}^r$ leads to an equivariant map

$$
f : (\mathbf{R}P^{2m-1}, i) \multimap (S^{r-1}, -).
$$

In particular we have $r(m) \geq s(m)$ for all m.

For $m = 1$ we find $s(1) \leq 2$, and indeed $s(1) = 2$, since $(\mathbf{R}P^1, i)$ is homeomorphic to $(S^1, -)$ and $s(S^1, -) = 2$. For $m = 2$ we get $s(2) \leq r(2) = 3$, and the map $\varphi = (\varphi_1, \varphi_2, \varphi_3)$ from Proposition 1.4 induces an equivariant map $f : \mathbf{R}P^3 \multimap S^2$, $[z] \mapsto \frac{q(z)}{\|q(z)\|}$ $(q = \operatorname{Im} \varphi)$.

This map f is related to (but not directly induced from) the well-known Hopf map $h : S^3 \to S^2$ which follows from the multiplication of complex numbers:

$$\mathbf{C}^2 \supset S^3 \ni (z_1, z_2) \text{ with } |z_1|^2 + |z_2|^2 = 1 \xmapsto{h} (2z_1\bar{z}_2, |z_1|^2 - |z_2|^2) \in S^2 \subset \mathbf{R}^3.$$

In the next section we shall improve the lower estimate $m + 1 \leq r(m)$ to the estimate $m + 1 \leq s(m) \leq r(m)$ for all m. Of course this requires the use of some results of algebraic topology since, after all, $s(m)$ is a topological invariant. Finally, the exact value of $s(m)$ has been computed by S. Stolz. This result will be given without proof.

What about the level of even-dimensional real projective spaces $\mathbf{R}P^{2m}$? Here it follows from the elements of algebraic topology that these spaces have no involution i without fixpoints or, in other words, $s(\mathbf{R}P^{2m}, i) = \infty$ for all $m \in \mathbf{N}_0$ and all involutions i on $\mathbf{R}P^{2m}$.

The most natural way to prove this goes as follows:

1) Compute the homology groups of $X := \mathbf{R}P^{2m}$:

$$H_q(X) = \begin{cases} \mathbf{Z} & q = 0, \\ \mathbf{Z}/2 & \text{for} \quad q \text{ odd, } 0 < q < 2m, \\ 0 & \text{all other } q. \end{cases}$$

2) Compute the Euler characteristic of X:

$$\chi(X) = \sum_{q=0}^{\dim X} (-1)^q \text{ rank } H_q(X) = (-1)^0 \cdot 1 = 1.$$

3) Note that for every finite CW-complex Y and every finite unramified covering $X \to Y$ with k sheets one has $\chi(X) = k \cdot \chi(Y)$.

4) Suppose that $X = \mathbf{R}P^{2m}$ has an involution i without fixpoints. Then X is a double covering of the quotient spaces $Y = X/i$. Hence $\chi(X) = 2\chi(Y)$ is an even number. But this contradicts the fact that $\chi(X) = 1$.

For proofs see e.g. [Bredon 1993, Ch. IV, §13 and §14].

We close this section with a Note on some loosely related results about polynomial maps from spheres to spheres.

1.14 Note.

(1) Let $m \geq n$ be integers such that $m \geq 2^t > n$ for some $t \in \mathbf{N}$. Then every *polynomial map* $p : S^m \to S^n$ is constant.

(2) For given m (resp. n) let $n = q(m)$ denote the least (resp. let $m = p(n)$ denote the greatest) positive integer such that there exists a *non-constant quadratic form*

$$q : S^m \to S^n.$$

Then $p(m) \geq m \geq q(m)$ for all m,

$$q(2^t) = 2^t, \; p(2^t - 1) = 2^t - 1 \text{ for all } t,$$

and the numbers $q(m)$, $p(n)$ can be recursively computed from these values and the Radon–Hurwitz function ϱ.

(1) is due to [Wood 1968], (2) is proved in a recent paper by [Yiu 1994]. The proofs use the construction of Hopf-like maps, some homotopy arguments and some of the results of Chapters 1 and 2 on sums of squares, multiplicative quadratic forms and bilinear composition of forms.

§2. The Level of $\mathbf{R}P^{2m-1}$

As announced in section 1 we now want to prove the lower estimate $s(m) \geq m + 1$ for the level $s(m) = s(\mathbf{R}P^{2m-1}, i)$:

2.1 Theorem. $s(m) \geq m + 1$ for all $m \in \mathbf{N}$.

The proof will closely follow the paper [Pfister–Stolz 1987]. It consists of several steps. The main idea is the following:

An equivariant map $f : (\mathbf{R}P^{2m-1}, i) \multimap S^{n-1}$ leads to a corresponding map $\bar{f} : \mathbf{R}P^{2m-1}/i \to \mathbf{R}P^{n-1}$ between the quotient spaces, $\bar{f}$ induces a homomorphism $\bar{f}^*$ in (complex) K-theory. This gives the lower estimate for $s(m)$ in the case where m is even. Finally, the case m odd is reduced to the case m even by taking the join of two equivariant maps.

More generally let $\mathbf{Z}/k$ denote the cyclic group of order $k \geq 1$. It operates without fixpoints on $S^{2m-1} \subset \mathbf{C}^m$ by sending the pair $(\ell \bmod k, x)$ to $e^{\frac{2\pi i}{k} \cdot \ell} x$ (for $x \in S^{2m-1}$). The orbit space of S^{2m-1} with respect to this $(\mathbf{Z}/k)$-action is the *lens space* $L^{2m-1}(k)$. In particular we have $\mathbf{R}P^{2m-1} = L^{2m-1}(2)$ and $\mathbf{R}P^{2m-1}/i = L^{2m-1}(4)$. All these spaces are compact.

For any compact topological space X let $K(X)$ denote the Grothendieck ring of (isomorphism classes of) finite-dimensional *complex* vector bundles over X. For the computation of $K(X_k)$ for the lens space $X_k := L^{2m-1}(k)$ two *line bundles* on X_k are particularly important:

- The *trivial* line bundle $X_k \times \mathbf{C} \searrow X_k$: its class in $K(X_k)$ will be denoted by 1. It is the unit element of the ring $K(X_k)$.

– The *canonical* line bundle $L_k \searrow X_k$ whose fibre $p^{-1}(x)$ is given by the complex line $\mathbf{C}z$ where $z \in \mathbf{C}^m \backslash 0$ is any preimage of $x \in X_k$ under the canonical projection map

$$\mathbf{C}^m \backslash 0 \longrightarrow S^{2m-1} \longrightarrow X_k.$$

The class of L_k in $K(X_k)$ is denoted by η_k.

Finally let $\sigma_k := \eta_k - 1$. Up to sign this is the so-called *Euler class* of the bundle L_k. We have

2.2 Proposition. As a ring

$$K(L^{2m-1}(k)) = \mathbf{Z}[\sigma_k]/(\sigma_k^m, 1 - (1 + \sigma_k)^k).$$

PROOF. See [Karoubi 1978, Ch. IV, §2, 2.12.]

Assume now that we have an equivariant map $f : (\mathbf{R}P^{2m-1}, i) \dashrightarrow S^{2t-1}$ where t is as small as possible. It induces a map $\bar{f} : L^{2m-1}(4) \to L^{2t-1}(2)$ of the orbit spaces and a ring homomorphism (in the opposite direction)

$$\bar{f}^* : K(L^{2t-1}(2)) \longrightarrow K(L^{2m-1}(4)).$$

2.3 Lemma. $\bar{f}^*(\eta_2) = \eta_4 \cdot \eta_4$.

PROOF. This is nearly trivial since multiplication in the Grothendieck ring is given by the tensor product of vector bundles and since it is easy to see that

$$
\begin{array}{ccc}
F : & L_4 \otimes L_4 & \to & L_2 \\
& \downarrow & & \downarrow \\
\bar{f} : & L^{2m-1}(4) & \to & L^{2t-1}(2)
\end{array}
$$

where $F : [x, c \otimes c'] \mapsto [\bar{f}(x), cc']$ is a well-defined bundle map covering $\bar{f}$.

The lemma implies $\bar{f}^*(\sigma_2) = \bar{f}^*(\eta_2 - 1) = \eta_4^2 - 1 = (1 + \sigma_4)^2 - 1$, hence the additive order of $(1 + \sigma_4)^2 - 1$ in $K(L^{2m-1}(4))$ is a divisor of the additive order of σ_2 in $K(L^{2t-1}(2))$.

We compute these orders:

2.4 Lemma.

(1) $(1 + \sigma_4)^2 - 1 \in \mathbf{Z}[\sigma_4]/(\sigma_4^m, 1 - (1 + \sigma_4)^4)$ has additive order $2^{[m/2]}$.

(2) $\sigma_2 \in \mathbf{Z}[\sigma_2]/(\sigma_2^t, 1 - (1 + \sigma_2)^2)$ has additive order 2^{t-1}.

PROOF.

(1) For evaluating the order of an element ε of $\mathbf{Z}[\sigma_4]/(\sigma_4^m, 1-(1+\sigma_4)^4)$ which is represented by a polynomial $p \in \mathbf{Z}[\sigma_4]$ with *vanishing* constant term we expand the quotient $p/(1 - (1 + \sigma_4)^4)$ as a power series $\sum_{i=0}^{\infty} a_i \sigma_4^i$, $a_i \in \mathbf{Q}$. Then p is contained in the ideal $(\sigma_4^m, 1 - (1 + \sigma_4)^4)$ if and only if $a_0, \ldots, a_{m-2} \in \mathbf{Z}$. (This is an easy exercise!) More generally it follows that the additive order of the element ε is the lowest common multiple of the denominators of $a_0, \ldots, a_{m-2}$.

For $p = (1 + \sigma_4)^2 - 1$ we obtain

$$
\begin{aligned}
\frac{(1 + \sigma_4)^2 - 1}{1 - (1 + \sigma_4)^4} &= \frac{-1}{1 + (1 + \sigma_4)^2} = \frac{-1}{2 + 2\sigma_4 + \sigma_4^2} \\
&= \frac{-(2 - 2\sigma_4 + \sigma_4^2)}{(2 + \sigma_4^2)^2 - 4\sigma_4^2} = \frac{1}{4}(-2 + 2\sigma_4 - \sigma_4^2)\frac{1}{1 + \frac{\sigma_4^4}{4}} \\
&= \frac{1}{4}(-2 + 2\sigma_4 - \sigma_4^2)\sum_{k=0}^{\infty}(-1)^k\frac{\sigma_4^{4k}}{2^{2k}}.
\end{aligned}
$$

Hence $a_i = 0$ for $i \equiv 3 \bmod 4$, and the denominator of a_i equals 2^{2k+1} for $i = 4k$ or $4k + 1$ and 2^{2k+2} for $i = 4k + 2$. This shows that the order of $(1 + \sigma_4)^2 - 1$ is 2^{2k+1} for $m - 2 = 4k$, $4k + 1$ and 2^{2k+2} for $m - 2 = 4k + 2$, $4k + 3$. Thus in all cases $\mathrm{ord}((1 + \sigma_4)^2 - 1) = 2^{[\frac{m}{2}]}$.

(2) Similarly we have

$$
\frac{\sigma_2}{1 - (1 + \sigma_2)^2} = \frac{-1}{2 + \sigma_2} = -\frac{1}{2}\sum_{k=0}^{\infty}(-1)^k(\frac{\sigma_2}{2})^k.
$$

Hence the order of σ_2 in $\mathbf{Z}[\sigma_2]/(\sigma_2^t, 1-(1+\sigma_2)^2)$ equals the lowest common multiple of the numbers 2^{k+1} for $0 \le k \le t - 2$ which is 2^{t-1}.

For even m Lemmata 2.3 and 2.4 will be enough to prove the lower estimate for $s(m)$. For odd m we need a further property of $s(m)$ which is completely analogous to Proposition 1.8 for the invariant $r(m)$:

2.5 Proposition. $s(m_1 + m_2) \le s(m_1) + s(m_2)$.

PROOF. Recall the definition of the *join* $X * Y$ of two topological spaces X, Y: $X * Y$ is the quotient space of $X \times [0,1] \times Y$ by the equivalence relation $(x, 0, y) \sim (x', 0, y)$, $(x, 1, y) \sim (x, 1, y')$ for all $x, x' \in X$, $y, y' \in Y$. Given two real vector-spaces V, W the join $S(V) * S(W)$ of the correponding spheres can be identified with $S(V \oplus W)$. Moreover, this identification is compatible with the $(\mathbf{Z}/k)$-action (given by multiplication by $e^{\frac{2\pi i}{k}}$) on the spheres if V and W have even dimension. (For $k = 2$ the dimensions of V, W are arbitrary.)

Let now $f_j : \mathbf{R}P^{2m_j-1} \multimap S^{s(m_j)-1}$ $(j = 1, 2)$ be $(\mathbf{Z}/2)$-equivariant maps. Denote by $\tilde{f}_j : S^{2m_j-1} \to S^{s(m_j)-1}$ the composition of f_j with the projection map from the sphere to projective space. These maps are $(\mathbf{Z}/4)$-equivariant with respect to the $(\mathbf{Z}/4)$-action given by multiplication by $i = \sqrt{-1}$ on the domain and multiplication by -1 on the range. Then also their join

$$\tilde{f}_1 * \tilde{f}_2 : S^{2(m_1+m_2)-1} = S^{2m_1-1} * S^{2m_2-1} \to S^{s(m_1)-1} * S^{s(m_2)-1} = S^{s(m_1)+s(m_2)-1}$$

is a $(\mathbf{Z}/4)$-equivariant map. Passing to the quotient we obtain an equivariant map
$\mathbf{R}P^{2(m_1+m_2)-1} \multimap S^{s(m_1)+s(m_2)-1}$ which proves our assertion.

The *Proof of Theorem 2.1.* This is now easy:

(1) If m is even then Lemmata 2.3 and 2.4 imply

$$2^{[\frac{m}{2}]} \mid 2^{t-1}, \quad \text{i.e. } m \leq 2t - 2.$$

By definition of t this gives $s(m) \geq 2t - 1 \geq m + 1$.

(2) If m is odd let us assume for a moment that $s(m) \leq m$. Then Proposition 2.5 implies $s(2m) \leq 2m$ which contradicts part (1) of the proof. Hence we have $s(m) \geq m + 1$ for all $m \in \mathbf{N}$.

The rest of this section is devoted to some comments on the complete solution of the level problem for $\mathbf{R}P^{2m-1}$ by S. Stolz. The full proof cannot be given here since it involves quite heavy machinery from algebraic topology which clearly lies beyond the scope of this book.

2.6 Theorem. [Stolz 1989] Let $m \geq 2$. Then

$$s(m) = \begin{cases} m+1 & m \equiv 0, 2, \\ m+2 & \text{for} \quad m \equiv 1, 3, 4, 5, 7 \bmod 8, \\ m+3 & m \equiv 6. \end{cases}$$

IDEA OF PROOF.

(1) For a topological space (X, i) with fixpoint-free involution i the $(\mathbf{Z}/2)$-action on X (by i) and $\mathbf{R}$ (by $r \mapsto -r$) induces the *real* line bundle $L = X \underset{\mathbf{Z}/2}{\times} \mathbf{R}$ over the quotient space $Y = X/i$ of (X, i). Similarly we have $nL = X \underset{\mathbf{Z}/2}{\times} \mathbf{R}^n$ for every $n \in \mathbf{N}$. With this notation we obtain the following reformulation of the level $s(X, i)$:

Let $f : X \multimap S^{n-1}$ be a $(\mathbf{Z}/2)$-equivariant map where $n \geq s(X, i)$. Then the induced $(\mathbf{Z}/2)$-equivariant map $\mathrm{id} \times f : X \to X \times S^{n-1} \subset X \times \mathbf{R}^n$ gives a nowhere vanishing section $\sigma : Y \to nL$ of the vector bundle nL

on Y. Conversely a nowhere vanishing section σ of nL can be normalized and gives rise to an equivariant $f : X \dashrightarrow S^{n-1}$. Hence $s(X, i)$ is the smallest number $n \in \mathbb{N}$ such that nL has a nowhere vanishing section σ.

An obstruction for the existence of such a section is the Euler class $e(nL)$. Euler classes of vector or other bundles exist in various generalized cohomology theories, e.g. in (co-)homology, (co-)homotopy, K-theory (complex vector bundles), KO-theory (real vector bundles) etc. For instance a study of the K-theory Euler class of L and nL would lead to the old estimate $s(m) \geq m + 1$ for $X = \mathbb{R}P^{2m-1} = L^{2m-1}(2)$, $Y = L^{2m-1}(4)$.

(2) Stolz investigates the *cohomotopy* Euler class of nL. This is more natural since the (non-)existence of an equivariant map $f : X \dashrightarrow S^{n-1}$ is a question of cohomotopy. More important is the fact that the vanishing of the cohomotopy Euler class of nL over Y is *necessary and sufficient* for the existence of a nowhere vanishing section σ of nL, hence for the existence of f, provided that the dimension of the bundle is large enough compared with the dimension of the base space. This results from

2.7 Proposition. (Crabb 1980, Prop. 2.4) Let α be an n-dimensional (real) vector bundle over a finite CW-complex Y. Suppose $\dim Y < 2(n - 1)$. Then α has a nowhere vanishing section if and only if its cohomotopy Euler class vanishes.

In our case we have $\dim Y = 2m - 1 < 2m$. Hence the condition on the dimensions is satisfied if $m \leq n - 1$. This holds for $n = s(m)$ since $s(m) \geq m + 1$ by Theorem 2.1.

(3) Let

$$s^*(m) = \begin{cases} m + 1 & m \equiv 0, 2, 4, \\ m + 2 & \text{for} \quad m \equiv 1, 3, 5, 7 \bmod 8 \quad (m \neq 1), \\ m + 3 & m \equiv 6. \end{cases}$$

Then it can be shown that the cohomotopy class of $nL \searrow Y$ does *not* vanish for $n = s^*(m) - 1$. Thus we get $s(m) \geq s^*(m)$. This is a small but important improvement of the old estimate $s(m) \geq m + 1$. The proof of the inequality $s(m) \geq s^*(m)$ for $m \geq 2$ uses various connections between the cohomology, cohomotopy, K-theory and KO-theory Euler classes as well as Gysin sequence, Hurewicz homomorphism, Hopf sphere bundles, Chern classes, Todd genus etc. together with some explicit calculations as in the proof of Lemma 2.4.

(4) Let

$$s^{**}(m) = \begin{cases} m + 1 & m \equiv 0, 2, \\ m + 2 & \text{for} \quad m \equiv 1, 3, 5, 7 \bmod 8, \\ m + 3 & m \equiv 4, 6. \end{cases}$$

Then it can be shown that the cohomotopy Euler class of nL vanishes for $n = s^{**}(m)$. In other words $s(m) \leq s^{**}(m)$. This part of Stolz' proof uses the Adams spectral sequence in an essential manner.

(5) Combining (3) and (4) we see that $s(m) = s^*(m) = s^{**}(m)$ for $m \geq 2$ unless $m \equiv 4 \bmod 8$. For $m \equiv 4 \bmod 8$ we get $m + 1 \leq s(m) \leq m + 3$. This last case can be handled by using Proposition 2.5:
We have $s(m) \leq s(m-2)+s(2) = m-1+3 = m+2$ since $m-2 \equiv 2 \bmod 8$, and $m + 5 = s(m + 2) \leq s(m) + s(2) = s(m) + 3$ since $m + 2 \equiv 6 \bmod 8$. Hence $s(m) = m + 2$ for $m \equiv 4 \bmod 8$ and the proof of Theorem 2.6 is finished.

2.8 Final comments.

(1) The proofs of 2.7 and hence of 2.6 are pure *existence* proofs. They do not lead to a *construction* of an equivariant map $f : \mathbf{R}P^{2m-1} \dashrightarrow S^{s(m)-1}$. In particular it is not clear whether f can be chosen to be a polynomial map or even a quadratic map. In this respect the upper estimate $s(m) \leq r(m) \leq \frac{1}{2}(3m+1)$ which results from 1.9 and 1.13 is still of interest since it gives an explicit quadratic map

$$g : \mathbf{R}P^{2m-1} \dashrightarrow S^{r-1} \quad \text{for} \quad r = \frac{1}{2}(3m + 1).$$

(2) Nevertheless it is tempting to ask the

Open Question. Do we have $s(m) = r(m)$ for all m?

§3. Estimates for the Level of Complex Projective Spaces

As far as I know the precise level of $\mathbf{C}P^n$ is not yet known. But the methods and results of sections 1 and 2 allow us to derive nontrivial lower and upper bounds for the level of $\mathbf{C}P^n$. At first we ask whether $\mathbf{C}P^n$ has an involution without fixpoints. We have

$$H_q(\mathbf{C}P^n) = \begin{cases} \mathbf{Z} & \text{for} \quad q = 0, 2, \ldots, 2n, \\ 0 & \text{otherwise.} \end{cases}$$

This implies $\chi(\mathbf{C}P^n) = n + 1$. A necessary condition for the existence of a fixpoint-free involution is therefore that $n = 2m-1$ is odd (as in the real case). In this case we have $\mathbf{C}^{2m} = \mathbf{H}^m$ where $\mathbf{H}$ is the skew-field of quaternions. Let $\{1, i, j, ij\}$ be the standard basis of $\mathbf{H}$ as a 4-dimensional vector-space over $\mathbf{R}$.

Left multiplication by the quaternion j acts on $H = C \oplus Cj$, hence on $S^{4m-1} \subset H^m$. This is a $(Z/4)$-action on S^{4m-1} since $j^2 = -1$, $j^4 = 1$. It induces a $(Z/2)$-action on the quotient space $CP^{2m-1} = C^{2m}\backslash 0 \big/ C^* = S^{4m-1}/S^1$ which we will also denote by j. Explicitly we have: Let $0 \neq z = (z_1, z_2, \ldots, z_{2m-1}, z_{2m}) \in C^{2m}$ be a representing vector for the point $[z] \in CP^{2m-1}$. z is identified with $(z_1 + z_2 j, \ldots, z_{2m-1} + z_{2m} j) \in H^m$. Since $jz_1 = \bar{z}_1 j$, $jz_2 j = -\bar{z}_2$ we get

$$jz = (-\bar{z}_2 + \bar{z}_1 j, \ldots, -\bar{z}_{2m} + \bar{z}_{2m-1} j) \in H^m.$$

Assume $[jz] = [z] \in CP^{2m-1}$, i.e. $jz = cz \in C^{2m}$ for some $c \in C^\times$. Then

$$-\bar{z}_2 = cz_1, \quad \bar{z}_1 = cz_2, \quad z_1 = \bar{c}\bar{z}_2, \quad -\bar{z}_2 = |c|^2 \bar{z}_2$$

and similar equations for the other components of z. Since $|c|^2 > 0$ we find $z_1 = z_2 = \ldots = z_{2m} = 0$, a contradiction. This shows that j is a fixpoint free involution on CP^{2m-1}.

3.1 Notation. The level of the projective space CP^{2m-1} with its natural involution j is denoted by

$$s_c(m) := s(CP^{2m-1}, j).$$

We want to estimate $s_c(m)$ for all $m \in N$. An upper bound for $s_c(m)$ can be obtained as follows:
Let

$$q(z) := (z_1\bar{z}_2, |z_1|^2 - |z_2|^2, \ldots, z_{2m-1}\bar{z}_{2m}, |z_{2m-1}|^2 - |z_{2m}|^2).$$

q is a map $C^{2m} \to (C \times R)^m = R^{3m}$ and is a quadratic map when considered as a map from R^{4m} to R^{3m}. We have $q(C^{2m}\backslash 0) \subset R^{3m}\backslash 0$ since $z_1\bar{z}_2 = 0 = |z_1|^2 - |z_2|^2$ implies $z_1 = z_2 = 0$ etc. Hence q induces a map $f : CP^{2m-1} \to S^{3m-1}$ defined by $f([z]) = \frac{q(z)}{\|q(z)\|}$.

Note that f is well-defined since $q(cz) = |c|^2 q(z)$ and $\|q(cz)\| = |c|^2\|q(z)\|$ for $c \in C^\times$. Finally, f is equivariant with respect to the involutions j on CP^{2m-1} and $-$ on S^{3m-1} since

$$j([z]) = [jz] \quad \text{and} \quad q(jz) = (-\bar{z}_2 z_1, |z_2|^2 - |z_1|^2, \ldots) = -q(z).$$

This shows

3.2 Proposition. $s_c(m) \leq 3m$.

In order to derive a lower estimate for $s_c(m)$ we use the fact that $CP^{2m-1} = S^{4m-1}/S^1$ is a quotient space of $RP^{4m-1} = S^{4m-1}/\pm$. By the definition of the involution j on CP^{2m-1} the projection $pr : S^{4m-1} \to CP^{2m-1}$ is compatible with the left multiplication m_j by the quaternion j on the sphere S^{4m-1} and the action of j on CP^{2m-1}. More generally let m_h denote the left multiplication on S^{4m-1} by a quaternion h of norm 1. For $h := \frac{1+ij}{\sqrt{2}}$ the identity $jh = hi$ shows that the map m_h on S^{4m-1} is $(\mathbf{Z}/4)$-equivariant with respect to the $(\mathbf{Z}/4)$-action given by m_i on the domain and by m_j on the range.

Assume now that we have an equivariant map

$$f : (\mathbf{C}P^{2m-1}, j) \multimap (S^{t-1}, -).$$

Then the composition

$$g = f \circ pr \circ m_h : S^{4m-1} \xrightarrow{m_h} S^{4m-1} \xrightarrow{pr} \mathbf{C}P^{2m-1} \xrightarrow{f} S^{t-1}$$

is $(\mathbf{Z}/4)$-equivariant with respect to the maps m_i on S^{4m-1} and the antipodal involution on S^{t-1} (since $f \circ pr \circ m_h \circ m_i = f \circ pr \circ m_j \circ m_h = f \circ j \circ pr \circ m_h = -f \circ pr \circ m_h$). Therefore g induces a $(\mathbf{Z}/2)$-equivariant map $\bar{g} : (\mathbf{R}P^{4m-1}, i) \multimap (S^{t-1}, -)$. This shows that we must have $t \geq s(2m)$. Taking the smallest possible value $t = s_c(m)$ Theorem 2.6 implies

3.3 Proposition.

$$s_c(m) \geq s(2m) = \begin{cases} 2m+1 & m \equiv 0, 1, \\ 2m+2 & \text{for} \quad m \equiv 2 \quad \mod 4, \\ 2m+3 & m \equiv 3. \end{cases}$$

The slightly weaker estimate $s_c(m) \geq 2m+1$ for all m follows of course already from Theorem 2.1.

Bibliography

I. Books on Quadratic Forms

[B]   R. Baeza: Quadratic Forms over Semilocal Rings. Lecture Notes in Math. **655**. Springer, Berlin 1978.

[C]   J.W.S. Cassels: Rational Quadratic Forms. Academic Press, London 1978.

[K]   M. Kneser: Quadratische Formen. Vorlesungsausarbeitung Göttingen (2. Aufl.) 1992.

[L]   T.Y. Lam: The Algebraic Theory of Quadratic Forms. Benjamin, Reading, Mass. 1973.

[Lo]   F. Lorenz, Quadratische Formen über Körpern. Lecture Notes in Math. **130**. Springer, Berlin 1970.

[MH]   J. Milnor – D. Husemoller: Symmetric Bilinear Forms. Springer, Berlin 1973.

[O'M]   O.T. O'Meara: Introduction to Quadratic Forms. Springer, Berlin 1963.

[R]   A.R. Rajwade: Squares. Cambridge Univ. Press 1993.

[S]   W. Scharlau: Quadratic and Hermitian Forms. Springer, Berlin 1985.

II. Other References

A.A. Albert 1942: Quadratic forms admitting composition. Annals of Math. **43**, 161–177 (1942).

Y. Alemu 1985: On zeros of forms over local fields. Acta Arithm. **45**, 163–171 (1985).

M. Amer 1976: Quadratische Formen über Funktionenkörpern. Dissertation Mainz 1976 (unpublished).

J.K. Arason 1975: Cohomologische Invarianten quadratischer Formen. J. Algebra **36**, 448–491 (1975).

J.K. Arason 1979: Wittring und Galoiskohomologie bei Charakteristik 2. J. reine angew. Math. **307/308**, 247–256 (1979).

J.K. Arason 1984: A proof of Merkurjev's theorem. Canad. Math. Soc. Conf. Proc. **4**, 121–130 (1984).

J.K. Arason – R. Elman – B. Jacob 1989: On quadratic forms and Galois cohomology. Rocky Mt. J. Math. **19**, 575–588 (1989).

J.K. Arason – A. Pfister 1971: Beweis des Krullschen Durchschnittsatzes für den Wittring. Invent. math. **12**, 173–176 (1971).

J.K. Arason – A. Pfister 1977: Zur Theorie der quadratischen Formen über formalreellen Körpern. Math. Z. **153**, 289–296 (1977).

J.K. Arason – A. Pfister 1982: Quadratische Formen über affinen Algebren und ein algebraischer Beweis des Satzes von Borsuk-Ulam. J. reine angew. Math. **331**, 181–184 (1982).

R. Aravire – R. Baeza 1989: The behavior of the v-invariant of a field of characteristic 2 under finite extensions. Rocky Mt. J. Math. **19**, 589–600 (1989).

C. Arf 1941: Untersuchungen über quadratische Formen in Körpern der Charakteristik 2 (Teil I). J. reine angew. Math. **183**, 148–167 (1941).

E. Artin 1927: Über die Zerlegung definiter Funktionen in Quadrate. Hamb. Abh. **5**, 100–115 (1927).

E. Artin – O. Schreier 1927: Algebraische Konstruktion reller Körper. Hamb. Abh. **5**, 85-99 (1927).

J. Ax – S. Kochen 1965: Diophantine problems over local fields I. Amer. J. of Math. **87**, 605-630 (1965).

R. Baeza 1982: Comparing u-invariants of fields of characteristic 2. Bol. Soc. Bras. Mat. **13**, 105-114 (1982).

F. Behrend 1940: Über Systeme reeller algebraischer Gleichungen. Compos. Math. **7**, 1-19 (1940).

B.J. Birch – D.J. Lewis 1965: Systems of three quadratic forms. Acta Arithm. **10**, 423-442 (1965).

B.J. Birch – D.J. Lewis – T.G. Murphy 1962: Simultaneous quadratic forms. Amer. J. Math. **84** , 110–116 (1962).

J. Bochnak – M. Coste – M.F. Roy 1987: Géométrie algébrique réelle. Springer Berlin 1987.

K. Borsuk 1933: Drei Sätze über die n-dimensionale euklidische Sphäre. Fund. Math. **20**, 177–190 (1933).

G.E. Bredon 1993: Topology and Geometry. Springer Berlin 1993.

L.E.J. Brouwer 1912: Über Abbildung von Mannigfaltigkeiten. Math. Ann. **71**, 97–115 (1912).

J. Browkin 1966: On forms over p-adic fields. Bull. Ac. Polon. Sci. **14**, 489–492 (1966).

A. Brumer 1978: Remarques sur les couples de formes quadratiques. C.R. Acad. Sci. Paris, Sér. A **286**, 679–681 (1978).

J.C. Burkill – H. Burkill 1970: A second course in mathematical analysis. Cambridge Univ. Press 1970.

J.W.S. Cassels 1964: On the representation of rational functions as sums of squares. Acta Arithm. **9**, 79–82 (1964).

J.W.S. Cassels 1979: On a problem of Pfister about systems of quadratic forms. Arch. Math. **33**, 29–32 (1979).

J.W.S. Cassels – W.J. Ellison – A. Pfister 1971: On sums of squares and on elliptic curves over function fields. J. Nb. Th. **3**, 125–149 (1971).

C. Chevalley 1936: Démonstration d'une hypothèse de M. Artin. Hamb. Abh. **11**, 73–75 (1936).

M.D. Choi – Z.D. Dai – T.Y. Lam – B. Reznick 1982: The Pythagoras number of some affine algebras and local algebras. J. reine angew. Math. **336**, 45–82 (1982).

M.C. Choi – T.Y. Lam – B. Reznick – A. Rosenberg 1980: Sums of squares in some integral domains. J. Algebra **65**, 234–256 (1980).

J.-L. Colliot-Thélène 1981: Variantes du Nullstellensatz réel et anneaux formellement réels. In: Géométrie Algébrique Réelle: Springer Lecture Notes in Math. **959**, 98–108 (1981).

J.-L. Colliot-Thélène 1993: The Noether-Lefschetz theorem and sums of 4 squares in the rational function field $R(x, y)$. Compos. Math. **86**, 235–243 (1993).

J.-L. Colliot-Thélène – U. Jannsen 1991: Sommes de carrés dans les corps de fonctions. C.R. Acad. Sci. Paris, Sér. I. **312**, 759–762 (1991).

J.-L. Colliot-Thélène – J.-J. Sansuc – P. Swinnerton-Dyer 1987: Intersections of two quadrics and Châtelet surfaces.
I: J. reine angew. Math. **373**, 37–107 (1987).
II: J. reine angew. Math. **374**, 72–168 (1987).

D. Coray 1980: On a problem of Pfister about intersections of three quadrics. Arch. Math. **34**, 403–411 (1980).

M.C. Crabb 1980: Z/2-Homotopy Theory. London Math. Soc. Lecture Notes **44**, Cambridge Univ. Press 1980.

Z.D. Dai – T.Y. Lam 1984: Levels in algebra and topology. Comment. Math. Helv. **59**, 376–424 (1984).

Z.D. Dai – T.Y. Lam – C.K. Peng 1980: Levels in algebra and topology. Bull. Amer. Math. Soc. **3**, 845–848 (1980).

V.B. Demjanov 1956: Pairs of quadratic forms over a complete field with a discrete norm with a finite residue class field. Izv. Akad. Nauk SSSR **20**, 307–324 (1956) (in Russian).

M. Denert – J.-P. Tignol – J. Van Geel – N. Vast 1990: The level of cyclic division algebras. Math. Z. **205**, 603–616 (1990).

D.W. Dubois 1967: Note on Artin's solution of Hilbert's 17th problem. Bull. Amer. Math. Soc. **73**, 540–541 (1967).

R. Elman 1977: Quadratic forms and the u-invariant III. Conf. on quadr. forms 1976. Queen's papers in pure and appl. math. No. **46**, 422–444 (1977).

R. Elman – T.Y. Lam 1973₁: Quadratic forms and the u-invariant I. Math. Z. **131**, 283–304 (1973).

R. Elman – T.Y. Lam 1973₂: Quadratic forms and the u-invariant II. Invent. math. **21**, 125–137 (1973).

R. Elman – T.Y. Lam 1976: Quadratic forms under algebraic extensions. Math. Ann. **219**, 21–42 (1976).

R. Elman – T.Y. Lam – A. Prestel 1973: On some Hasse principles over formally real fields. Math. Z. **134**, 291–301 (1973).

R. Elman – A. Prestel 1984: Reduced stability of the Witt ring of a field and its pythagorean closure. Amer. J. Math. **106**, 1237–1260 (1984).

W. Fulton 1984: Intersection Theory. Springer, Berlin 1984.

M. Greenberg 1966: Rational points in henselian discrete valuation rings. Publ. Math. IHES **31**, 59–64 (1966).

M. Greenberg 1969: Lectures on forms in many variables. Benjamin, New York 1969.

Z. Guangxin 1988: A characterization of preordered fields with the weak Hilbert property. Proc. Amer. Math. Soc. **104**, 335–342 (1988).

Z. Guangxin 1991: On preordered fields related to Hilbert's 17th problem. Math. Z. **206**, 145–151 (1991).

W. Habicht 1940: Über die Zerlegung strikte definiter Formen in Quadrate. Comment. Math. Helv. **12**, 317–322 (1940).

D. Hilbert 1888: Über die Darstellung definiter Formen als Summe von Formenquadraten. Math. Ann. **32**, 342–350 (1888) = Ges. Abh. II, 154-161.

D. Hilbert 1893_1: Über ternäre definite Formen. Acta Math. **17**, 169-198 (1893) = Ges. Abh. II, 345–366.

D. Hilbert 1893_2: Über die vollen Invariantensysteme. Math. Ann. **42**, 313-373 (1893) = Ges. Abh. II, 287–344.

D. Hilbert 1906: Mathematische Probleme. Göttinger Nachr. 1906, 253–297 = Ges. Abh. III, 290–329.

K. Hoffman – R. Kunze 1971: Linear Algebra (2nd ed.) Prentice-Hall, Englewood Cliffs N.J. 1971.

H. Hopf 1940: Ein topologischer Beitrag zur reellen Algebra. Comment. Math. Helv. **13**, 219–239 (1940/41).

E.A.M. Hornix 1991: Formally real fields with prescribed invariants in the theory of quadratic forms. Indag. Math. **2**, 65–78 (1991).

J.S. Hsia - R. Johnson 1974: On the representation in sums of squares for definite functions in one variable over an algebraic number field. Amer. J. Math. **96**, 448–453 (1974).

A. Hurwitz 1923: Über die Komposition der quadratischen Formen. Math. Ann. **88**, 1–25 (1923, posthum.).

B. Jacob – M. Rost 1989: Degree four cohomology invariants for quadratic forms. Invent. math. **96**, 551–570 (1989).

N. Jacobson 1980: Basic Algebra, Vol. 2. Freeman, San Francisco 1980.

I. Kaplansky 1953: Quadratic forms. J. Math. Soc. Japan **6**, 200–207 (1953).

M. Karoubi 1978: K-Theory. An Introduction. Springer, Berlin 1978.

K. Kato 1982: Symmetric bilinear forms, quadratic forms and Milnor K-theory in characteristic two. Invent. math. **66**, 493–510 (1982).

K. Kato 1986: A Hasse principle for two-dimensional global fields. J. reine angew. Math. **336**, 142–183 (1986).

I. Kersten 1990: Brauergruppen von Körpern. Vieweg, Braunschweig 1990.

M. Knebusch 1982: An algebraic proof of the Borsuk-Ulam theorem for polynomial mappings. Proc. Amer. Math. Soc. **84**, 29–32 (1982).

M. Knebusch – M. Kolster 1982: Wittrings. Aspects of Mathematics, vol E2. Vieweg, Braunschweig–Wiesbaden 1982.

M. Knebusch – W. Scharlau 1980: Algebraic theory of quadratic forms. Generic methods and Pfister forms. (DMV-Seminar 1, Notes by Heisook Lee), Birkhäuser, Basel 1980.

H. Kneser 1934: Verschwindende Quadratsummen in Körpern. Jber. Dt. Math. Ver. **44**, 143–146 (1934).

M. Kneser 1978: Konstruktive Lösung p-adischer Gleichungssysteme. Nachr. Akad. Wiss. Göttingen. Math.-Phys. Kl. Jg. 1978, Nr. 5

G. Kreisel 1957: Hilbert's 17th problem. Summaries of talks presented at the Summer Inst. of Symbolic Logic at Cornell Univ., 313–320 (1957).

T.Y. Lam 1989_1: Fields of u-invariant 6 after A. Merkurjev. P. 12–30 in: Ring Theory 1989 in Honor of S. Amitsur (ed. L. Rowen), Weizmann Science Press, Jerusalem 1989.

T.Y. Lam 1989_2: Some consequences of Merkurjev's work on function fields. Preprint 1989.

E. Landau 1906: Über die Darstellung definiter Funktionen durch Quadrate. Math. Ann. **62**, 272–285 (1906). = Coll. Works, vol 2, no. 38.

S. Lang 1952: On quasi-algebraic closure. Ann. of Math. **55**, 373–390 (1952).

S. Lang 1953: The theory of real places. Ann. of Math. **57**, 378–391 (1953).

S. Lang 1965: Algebra. Addison-Wesley, Reading/Mass. 1965.

S. Lang 1991: Number Theory III. Encycl. of Math. Sciences, vol 60. Springer, Berlin 1991.

D.B. Leep 1984: Systems of quadratic forms. J. reine angew. Math. **350**, 109–116 (1984).

D.B. Leep 1988: Invariants of quadratic forms under algebraic extensions. Preprint 1988.

D.B. Leep 1990_1: Pfister's conjecture on quadratic C_0-fields. J. reine angew. Math. **404**, 209–220 (1990).

D.B. Leep 1990_2: Levels of division algebras. Glasgow Math. J. **32**, 365–370 (1990).

D.B. Leep – A.S. Merkurjev 1994: Growth of the u-invariant under algebraic extensions. Contemp. Math. **155**, 327–332 (1994).

D.B. Leep – J.-P. Tignol – N. Vast 1989: The level of divison algebras over local and global fields. J. Nb. Th. **33**, 53–70 (1989).

D.W. Lewis 1987: On the level. Irish Math. Soc. Bull. **19**, 33–48 (1987).

F. Lorenz 1990: Einführung in die Algebra, Teil II. BI Wissenschaftsverlag, Mannheim 1990.

L. Mahé 1990: Level and Pythagoras number of some geometric rings. Math. Z. **204**, 615–629 (1990).

L. Mahé 1992: Erratum. Math. Z. **209**, 481–483 (1992).

P. Mammone – R. Moresi – A.R. Wadsworth 1991: u-invariants of fields of characteristic 2. Math. Z. **208**, 335–347 (1991).

P. Mammone – J.-P. Tignol – A.R. Wadsworth 1991: Fields of characteristic 2 with prescribed u-invariants. Math. Ann. **290**, 109–128 (1991).

K. McKenna 1975: New facts about Hilbert's seventeenth problem. Springer Lecture Notes in Math. **498**, 220–230 (1975).

A.S. Merkurjev 1981: On the norm residue symbol of degree 2. Soviet Math. Doklady **24**, 546–551 (1981).

A.S. Merkurjev 1992: Simple algebras and quadratic forms. Math. USSR Isvestija **38**, 215–221 (1992).

A. Merkurjev – A. Suslin 1990: The norm residue homomorphism of degree 3. Math. USSR Isvestija **36**, 339–356 (1990).

A. Merkurjev – A. Suslin 1991: The group K_3 for a field. Math. USSR Isvestija **36**, 541–565 (1991).

J. Milnor 1970: Algebraic K-theory and quadratic forms. Invent. math. **9**, 318–344 (1970).

J. Milnor 1971: Symmetric inner product spaces in characteristic two. Ann. of Math. Studies **70**. Princeton Univ. Press 1971.

J. Mináč 1994: Remarks on Merkurjev's investigations of the u-invariant. Contemp. Math. **155**, 333–338 (1994).

J. Mináč – A.R. Wadsworth 1993: The u-invariant for algebraic extensions (Preprint 1993).

T.S. Motzkin 1967: The arithmetic-geometric inequalitiy. In: Proc. Symp. on Inequalities (ed. O. Shisha), 205–224. Acad. Press, New York 1967.

M. Peters 1972: Die Stufe von Ordnungen ganzer Zahlen in algebraischen Zahlkörpern. Math. Ann. **195**, 309–314 (1972).

M. Peters 1974: Summen von Quadraten in Zahlringen. J. reine angew. Math. **268/269**, 318–323 (1974).

A. Pfister 1965_1: Darstellung von −1 als Summe von Quadraten in einem Körper. J. London Math. Soc. **40**, 159–165 (1965).

A. Pfister 1965_2: Multiplikative quadratische Formen. Arch. Math. **16**, 363–370 (1965).

A. Pfister 1966: Quadratische Formen in beliebigen Körpern. Invent. math. **1**, 116–132 (1966).

A. Pfister 1967_1: Lectures on "Quadratic Forms" (Notes by A.D. M^{C}Gettrick). Department of Pure Mathematics and Mathematical Statistics, Cambridge GB 1967.

A. Pfister 1967_2: Zur Darstellung definiter Funktionen als Summe von Quadraten. Invent. math. **4**, 229–237 (1967).

A. Pfister 1971_1: Quadratic Forms over Fields. Proc. Sympos. Pure Math. Vol XX, 150–160. Amer. Math. Soc., Providence, R.I. 1971.

A. Pfister 1971_2: Sums of squares in real function fields. Actes du Congrès Internat. des Math. (Nice 1970), Tome I, 297–300. Gauthier-Villars, Paris 1971.

A. Pfister 1976: Hilbert's seventeenth problem and related problems on definite forms. Proc. Sympos. Pure Math. Vol XXVIII, 483–489. Amer. Math. Soc., Providence, R.I. 1976.

A. Pfister 1979: Systems of quadratic forms. Bull. Soc. Math. France, Mémoire **59**, 115–123 (1979).

A. Pfister 1982: On quadratic forms and abelian varieties over function fields. Contemp. Math. **8**, 249–264 (1982).

A. Pfister 1984: Systems of quadratic forms II. Rocky Mt. J. Math. **14**, 973–976 (1984).

A. Pfister 1989: Systeme quadratischer Formen III. J. reine angew. Math. **394**, 208–220 (1989).

A. Pfister – S. Stolz 1987: On the level of projective spaces. Comment. Math. Helv. **62**, 286–291 (1987).

F. Pop 1991: Summen von Quadraten. Preprint 1991.

Y. Pourchet 1971: Sur la représentation en somme de carrés des polynômes à une indéterminée sur un corps de nombres algébriques. Acta Arithm. **19**, 89–104 (1971).

A. Prestel 1978: Remarks on the Pythagoras and Hasse number of real fields. J. reine angew. Math. **303/304**, 284–294 (1978).

C. Riehm 1964: On the integral representation of quadratic forms. Amer. J. Math. **86**, 25–62 (1964).

A. Robinson 1955: On ordered fields and definite functions. Math. Ann. **130**, 257–271 (1955).

I.R. Šafarevič 1965: Algebraic Surfaces. Proc. Steklov Inst. of Math. **7** (1965).

C.-H. Sah 1962: Symmetric bilinear forms and quadratic forms. J. Algebra **20**, 144–160 (1962).

T. Sander 1991: Existence and uniqueness of the real closure of an ordered field without Zorn's Lemma. J. Pure Appl. Algebra **73**, 165–180 (1991).

R. Scharlau 1980: On the Pythagoras number of orders in totally real number fields. J. reine angew. Math. **316**, 208–210 (1980).

D.B. Shapiro 1984: Products of sums of squares. Expos. Math. **2**, 235–261 (1984).

D.B. Shapiro: Composition of Quadratic Forms. W. de Gruyter.

C.L. Siegel 1921: Darstellung total positiver Zahlen durch Quadrate. Math. Z. **11**, 246–275 (1921) = Ges. Abh. I, 47–76.

T.A. Springer 1955: Quadratic forms over a field with a discrete valuation. Indag. Math. **17**, 352–362 (1955).

G. Stengle 1979: Integral solution of Hilbert's seventeenth problem. Math. Ann. **246**, 33–39 (1979).

S. Stolz 1989: The level of real projective spaces. Comment. Math. Helv. **64**, 661–674 (1989).

G. Terjanian 1966: Un contre-example à une conjecture d'Artin. C.R. Acad. Sci. Paris, Sér. A **262**, 612 (1966).

G. Terjanian 1967: Sur la dimension diophantienne des corps p-adiques. Acta Arithm. **34**, 127–130 (1967).

G. Terjanian 1972: Dimension arithmétique d'un corps. J. Algebra **22**, 517–545 (1972).

J.-P. Tignol 1990: Réduction de l'indice d'une algèbre simple centrale sur le corps de fonctions d'une quadrique. Bull. Soc. Math. Belg. **42**, 735–745 (1990).

C. Tsen 1933: Divisionsalgebren über Funktionenkörpern. Nachr. Wiss. Ges. zu Göttingen (Math.-Ph. Kl.) 1933, 335–339.

C. Tsen 1936: Zur Stufentheorie der Quasi-algebraisch-Abgeschlossenheit kommutativer Körper. J. Chinese Math. Soc. **1**, 81–92 (1936).

E. Witt 1934: Zerlegung reeller algebraischer Funktionen in Quadrate, Schiefkörper über reellem Funktionenkörper. J. reine angew. Math. **171**, 4–11 (1934).

E. Witt 1937: Theorie der quadratischen Formen in beliebigen Körpern. J. reine angew. Math. **176**, 31–44 (1937).

R. Wood 1968: Polynomial maps from spheres to spheres. Invent. math. **5**, 163–168 (1968).

P.Yiu 1994: Quadratic forms between euclidean spheres. Manuscripta math. **83**, 171–181 (1994).

List of Symbols

N, Z, Q, R, C, H:
Sets of natural numbers, integers, rationals, real numbers, complex numbers, Hamilton quaternions

N_0 = $\{0\} \cup N$
F_q :   finite field with q elements (for a prime power q)
Q_p : p-adic field (for a prime number p)
$|S|$:   number of elements of the finite set S

Index

Printed in the United States
By Bookmasters